America's New Slavery?

Behind the Scenes and Updates

Or America's Salvation?

All Should Be Concerned!

America's New Slavery?

Behind the Scenes and Updates

Or America's Salvation?

All Should Be Concerned!
_____

Jose Collazo

To order additional copies of this book, contact:
Xlibris Corporation
1-888-795-4274
www.Xlibris.com
Orders@Xlibris.com
91447

CONTENTS

Dedication

To my Lord and God, with great love and devotion, I answer the call.

To my two daughters and granddaughter, may God bless and protect you always.

To all those who have helped me in this cause, God bless.

Introduction

I have found much more proof of F**unctional Magnetic Resonance Imaging** (FMRI) technology. This technology allows the U.S. government to read a person's thoughts (remotely). Many questions must be asked on how this technology is being used. How come September 11 was not prevented.

One hundred and eighty-three documents were withheld from the first book. In Roman v. DARPA, one hundred and eighty-three documents were sent out in December 2009, and I did not receive them until May 2010. I received them a couple of days after the first book was released.

In these documents, the **National Reconnaissance Office** (NRO) is proven to be lying in its claims that they have no records on FMRI technology. HAARP activist concerns on implanting thoughts are supported also. I update everyone on my five FOIA cases and snail-type justice.

I present new investigations on proof of the existence of God, tracking any person in the world by electromagnetic signatures, space travel, making objects move using some sort of electromagnetic technology.

I present the idea of a civil commission. This commission should have the authority to view all government records and report to the people. It should be able to hold hearings and question witnesses. It should be able to order criminal investigations. Justice is being delayed by our politicians and special-interest groups. For fourteen years I have been warning about FMRI technology and no one steps up to investigate.

My Mother's Worries about My Cause

My mother worries about my cause.

Should I stop, run, or hide?

Who else will answer freedom's call?

"No, No, Mother, this is my plight."

To defend our freedom until help is in sight.

To defend our freedom is my cause.

Even if I have to give my life.

Let it be, whatever my plight.

Let freedom ring!

Let freedom ring throughout the rest of our history.

Jose Collazo,
September 2010

These pages where just released by the NRO. Almost to late to put in this book.

US DISTRICT COURT
EASTERN DISTRICT OF NY

ROMAN, Plaintiff,

                                    NEW EVIDENCE TO BE ENTERED

V.   CASE NO. 09-2504

NRO, Defendants,

I have been waiting 13 months for FOIA tasking sheets from my July 1, 2009 and Oct. 15, 2009 request to the NRO. Again only one level of security was searched and no top level of security was searched. No copies of any duty officers forms authorizing searches. No copies of classified documents receipts authorizing searches. Emails have been released, in these emails:
1. Someone states the NRO is credited for MRI development.
2. Someone states information was found, is this true?
3. Someone states they plan not to answer.
4. Someone They never came up with anything concrete... just possible involvements. Going by our interpretations of what constitutes a "reasonable search"... and answered in AS & Ts 2nd search in response to the appeal they said they may have uncovered some info, Did They?
5. Multiple pages are blacked out and ten documents, consisting of tirty-seven pages are denied in full under exemptions. The court can review these documents in chamber.
6. Another really dated memory. T he NRO sponsored R & D in the early 90's in superconductivity. The link may be that research's relationship to the superconducting magnets in an MRI.
7. Warning: this document may not be used as a source of derivative classification.
8. We are under the gun to produce a response for General Counsel in this FOIA Appeal. This search should be SIMPLE!!! We need your response TODAY... Mr. Moffett has directed me to contact the Directorate of Security ih those offices that have not responded or he will call the directorate head...I really do not want that to happen, so please help us out. (Note SIMPLE SEARCH should be made)
9. IMINT made searches for MRI and magnetic resonance imaging and no records were found. (They must not be searching or looking in all the wrong places. Reason being they invented MRI technology)
10. We think we've at least met the threshold for definition of a reasonable search.
11. Superconductivity Project no. was RT=Reconnaissance Technology, 23= Electronics; Program C (Navy); 009= an assigned project number. (Remember the book Spies, Lies, and Whistleblowers on pages 115-116. The two authors, two ex British agents stated: Files were hidden in a two step fold, one you needed a code word RT=Reconnaissance Technology and a # 23=Electronics and if this sequence was not followed, you would get a no records response from the computer. They have hidden over a million files on British citizens.)
12. Mr. Barlow did not sign the F09-0064 appeal letter. He asked the IG to reconsider redacting portions of the responsive documents.

Cc. US DISTRICT COURT   Robert Kambic
      EASTERN DISTRICT OF NY   US Attorney's office
      100 Federal plaza   610 Federal plaza
      Central Islip, NY 11722   Central Islip, NY 11722

I swear the foregoing is true and correct under penalty of perjury.   Gilbert Roman
                                                 PO Box 170109
                                                 Ozone Pk., NY 11417

NATIONAL RECONNAISSANCE OFFICE
14675 Lee Road
Chantilly, VA 20151-1715

23 November 2010

Gilbert Roman
P.O. Box 170109
Ozone Park, NY 11417

Dear Mr. Roman:

This is in response to your letter dated 22 October 2009, received in the Information Management Services Center of the National Reconnaissance Office (NRO) on 30 October 2009. Pursuant to the provisions of the Freedom of Information Act (FOIA), inclusive of the Privacy Act (PA), you are asking for:

"1...copies of all the Freedom of Information and/or Privacy Act of 1974 task sheets used to process my request to your agency; which you responded to on July 1, 2009 and Oct. 15, 2009...;

2. Copies of the DUTY OFFICER forms authorizing these searches...;

3. Copies of forms from the CLASSIFIED DOCUMENTS RECEIPTS...;

4. Copies of the forms from the OFFICE OF THE CLASSIFIED REGISTER OF CONTROL...;

5. Copies of any and all memorandums, emails concerning Gilbert Roman (ME)."

Your request was processed in accordance with the PA and the FOIA, 5 U.S.C. § 552, as amended. A thorough search of our records and databases located forty-one documents consisting of 449 pages responsive to your request. Ten documents, consisting of thirty-seven pages are denied in full. The remaining thirty-one documents, totaling 412 pages, are being released to you in part.

The material that is denied is withheld pursuant to:

FOIA exemption (b)(3) 10 U.S.C. § 424 which states: "Except as required by the President or as provided in subsection (c), no provision of law shall be construed to require the disclosure of (1) The organization or any function... (2)... number of persons employed by or assigned or detailed to any such organization or the name, official title, occupational series, grade, or salary of any such person... (b) Covered Organizations... the National Reconnaissance Office";

FOIA exemption (b)(6) which apply to records which, if released, would constitute a clearly unwarranted invasion of the personal privacy of individuals, and 5 U.S.C. §552a(b), which concerns information about other individuals which may not be released without their written consent;

FOIA exemption (b)(5), which exempts deliberative, attorney work product and attorney-client privilege information, and Privacy Act exemption (d)(5) which applies to information compiled in reasonable anticipation of a civil action proceeding.

You have the right to appeal this determination by addressing your appeal to the NRO FOIA/PA Appeal Authority, 14675 Lee Road, Chantilly, VA 20151-1715, within 60 days of the date of this letter. Should you decide to do so, please explain the basis of your appeal.

If you have any questions, please call the Requester Service Center at 703-227-9326, and reference case numbers P10-0079/F10-0034.

Sincerely,

Stephen R. Glenn
Chief, Information Access
and Release Team

Attachments: Responsive documents from case files, F09-0063

(S//xx) Another really dated memory. The NRO sponsored R&D in the early 90's in superconductivity. The link may be that research's relationship to the superconducting magnets in an MRI.

b3 b6

From:
Sent: Tuesday, September 01, 2009 2:00 PM   b3 b6
To: b3 b6
Subject: RE: PRIORITY -- FOIA Case #F09-0063 -APPEAL (Gilbert Roman) -- SECRET//TALENT KEYHOLE//25X1 20590901 RRG dated July 2005

classification: SECRET//TALENT KEYHOLE//25X1 20590901 RRG dated July 2005

b3
b6

(S//xx) NRO R&D has actually been credited with inventing the MRI. Before my time and I have no further knowledge, but I do know in the early 90's we included it on charts for transition of NRO technology to the civil/commercial world. This type of request should be forwarded to the NRO archives, but I'm not aware of the procedure..

b3
b6

From: b3 b6
Sent: Tuesday, September 01, 2009 1:54 PM
To: ast   all b3
Subject: FW: PRIORITY -- FOIA Case #F09-0063 -APPEAL (Gilbert Roman) -- UNCLASSIFIED//FOUO
Importance: High

classification: UNCLASSIFIED//FOUO

(U) Tis apparently the season for FOIAs. I'm conducting a search of our file system "functional magnetic resonance imaging". If you have any documents or non-humorous responses please send them to me for consolidation as soon as possible. Otherwise I will negative response.

b3
b6

From: b3 b6
Sent: Tuesday, September 01, 2009 1:19 PM   b3 b6
To:
Subject: FW: PRIORITY -- FOIA Case #F09-0063 -APPEAL (Gilbert Roman) -- UNCLASSIFIED//FOUO
Importance: High

classification: UNCLASSIFIED//FOUO

Here is another FOIA request. This one fell through the cracks and now we're late. Request a quick search for "...information on functional magnetic resonance imaging...."

Thanks,

From:

Sent: Tuesday, September 01, 2009 6:45 AM

To: SE-DAG-INBOX

Cc: ast

Subject: FW: PRIORITY – FOIA Case #F09-0063 -APPEAL (Gilbert Roman) — UNCLASSIFIED//FOUO

Importance: High

classification: UNCLASSIFIED//FOUO

Good morning all.

(U//FOUO) Forgive me if I've missed something in the flood of emails, but I don't see AS&T and SE responses on this appeal request. Can you provide an update on the status? This is an appeal case and I need to move it forward as soon as possible to try to avoid litigation.

(U//FOUO) Any questions or concerns about this request, please contact me.

Thanks.

Senior Case Analyst
I-SC/ART

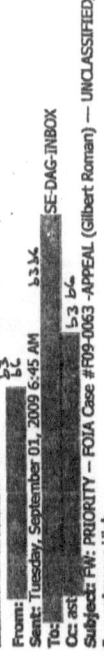

From:

Sent: Wednesday, July 22, 2009 1:05 PM

To:

UNCLASSIFIED//FOUO

UNCLASSIFIED//FOUO

UNCLASSIFIED//FOUO

UNCLASSIFIED//FOUO

DECL ON: 25X1 20590901 RRG dated July 2005
Derived From: NCG 6.0, 21 May 2005
SECRET//TALENT KEYHOLE//25X1 20590901 RRG dated July 2005

Warning: This document may not be used as a source of derivative classification.
DECL ON: 25X1 20590901 RRG dated July 2005
Derived From: NCG 6.0, 21 May 2005
SECRET//TALENT KEYHOLE//25X1 20590901 RRG dated July 2005

Warning: This document may not be used as a source of derivative classification.
DECL ON: 25X1 20590901 RRG dated July 2005
Derived From: NCG 6.0, 21 May 2005
SECRET//TALENT KEYHOLE//25X1 20590901 RRG dated July 2005

Warning: This document may not be used as a source of derivative classification.
DECL ON: 25X1 20590901 RRG dated July 2005
Derived From: NCG 6.0, 21 May 2005
SECRET//TALENT KEYHOLE//25X1 20590901 RRG dated July 2005

DECL ON: 25X1 20590901 RRG dated July 2005
Derived From: NCG 6.0, 21 May 2005
SECRET//TALENT KEYHOLE//25X1 20590901 RRG dated July 2005

Warning: This document may not be used as a source of derivative classification.
DECL ON: 25X1 20590901 RRG dated July 2005
Derived From: NCG 6.0, 21 May 2005
SECRET//TALENT KEYHOLE//25X1 20590901 RRG dated July 2005

9/16/2009

b6
(b)(3)

From: b6 b)(3)
Sent: Friday, September 25, 2009 1:45 PM
To: b6 b)(3)
Cc: FOIA (IART)
Subject: RE: F09-0063 — UNCLASSIFIED//FOUO
Classification: UNCLASSIFIED//FOUO

classification: UNCLASSIFIED//FOUO

They never came up with anything concrete.... just recollections of possible involvements. Going by our interpretations of what constitutes a "reasonable search," we decided that the search was complete, with no records.

b6
(b)(3)

b6 (b)(3)
Senior Case Analyst
IMSC/IART
(b)(3)

---

b6
From: (b)(3)
Sent: Friday, September 25, 2009 1:28 PM
To: (b)(3) b6
Subject: F09-0063 — UNCLASSIFIED//FOUO

classification: UNCLASSIFIED//FOUO

(b)(3) b6

(U//FOUO) In AS&Ts 2nd search in response to the appeal they said they may have uncovered some information. Did they?

Thanks.

(b)(3) b6

UNCLASSIFIED//FOUO

UNCLASSIFIED//FOUO

9/25/2009

From: b3 b6

Sent: Tuesday, September 01, 2009 12:50 PM

To: b3 b6 SE-DAG-INBOX b3 b6 ast-dag

Cc: FOIA (IART) b3 b6

Subject: RE: Gilbert Roman and F09-0063 Appeal/litigation — UNCLASSIFIED//FOUO

Classification: UNCLASSIFIED//FOUO

classification: UNCLASSIFIED//FOUO

I was on leave when responses came back from our SE components, and I lost the bubble on this in the meantime. SE has no records that are responsive to Mr. Roman's request.

b3
b6

b3
b6

From: b3 b6

Sent: Tuesday, September 01, 2009 10:27 AM

To: b3 b6 SE-DAG-INBOX; b3 b6 ast-dag

Cc: FOIA (IART)

Subject: FW: Gilbert Roman and F09-0063 Appeal/litigation — UNCLASSIFIED//FOUO

classification: UNCLASSIFIED//FOUO

We are under the gun to produce a response for General Counsel in this FOIA Appeal. This search should be SIMPLE!!! We need your response TODAY . . . Mr. Moffett has directed me to contact the Director of Security in those offices that have not responded or he will personally call the Directorate head . . . I really don't want that to happen, so please help us out.

This gentleman is currently litigating for our failure to respond . . . push this one to the top of the list and let's get it taken care of today, please. THANK YOU!

b3 b6

Lead b3 b6

FOIA/Privacy/Prepub
FESO/ASG/IMSC/IART
b3
Don't Forget to Tell Us How We're Doing...
Click Here to Open an
ASG/IMSC CUSTOMER SURVEY
We Look Forward to Hearing From You

From: [redacted] **b3 b6**
Sent: Tuesday, September 01, 2009 6:51 AM
To: [redacted] **b3 b6**
Cc: FOIA (IART)
Subject: Gilbert Roman and F09-0063 Appeal/litigation — UNCLASSIFIED//~~FOUO~~

classification: UNCLASSIFIED//~~FOUO~~

b3 b6

(U//FOUO) Good morning [redacted] This is just a brief update on where we stand on the processing of Mr. Gilbert's appeal of our "no records" response in F09-0063. I'm still waiting for a search/review response from a couple of the tasked De/Os. So far, IMINT, DDMS, and SIGINT have located no documents responsive to his request for "information on functional magnetic resonance imaging." I'm trying to corral the remaining responses so I can move a final package forward.

b3
b6

[redacted] **b3 b6**
Senior Case Analyst
FESC/IART **b3**

UNCLASSIFIED//~~FOUO~~

UNCLASSIFIED//~~FOUO~~

UNCLASSIFIED//~~FOUO~~

9/16/2009

xxi

Mr. Gilbert Roman
95-25 77th Street
Ozone Park, NY 11416

Dear Mr. Roman:

This is in response to your letter dated 12 July 2009, received in the Information Access and Release Center of the National Reconnaissance Office (NRO) on 21 July 2009, appealing our 1 July 2009 determination that the NRO has no records responsive to your 14 May 2009 request pursuant to the Freedom of Information Act for:

"1. ...information on functional magnetic resonance
        imaging.
2. The date it was put into service.
3. The first successful report on the first person it was
        used on successfully."

As the Appellate Authority, and after a complete review, I have determined that there are no National Reconnaissance Office (NRO) records responsive to your request. We have conducted reasonable searches of those components within the NRO that might have records within the parameters of your request. This action is taken in accordance with the provisions of 5 U.S.C. § 552.

You are advised that you are entitled to a judicial review of this determination in a United States District Court in accordance with 5 U.S.C. § 552.

Sincerely,

Charles Barlow

NATIONAL RECONNAISSANCE OFFICE
14675 Lee Road
Chantilly, VA 20151-1715

15 October 2009

Mr. Gilbert Roman
95-25 77th Street
Ozone Park, NY 11416

Dear Mr. Roman:

This is in response to your letter dated 12 July 2009, received in the Information Access and Release Center of the National Reconnaissance Office (NRO) on 21 July 2009, appealing our 1 July 2009 determination that the NRO has no records responsive to your 14 May 2009 request pursuant to the Freedom of Information Act for:

"1. ...information on functional magnetic resonance imaging.
2. The date it was put into service.
3. The first successful report on the first person it was used on successfully."

As the Appellate Authority, and after a complete review, I have determined that there are no National Reconnaissance Office (NRO) records responsive to your request. We have conducted reasonable searches of those components within the NRO that might have records within the parameters of your request. This action is taken in accordance with the provisions of 5 U.S.C. § 552.

You are advised that you are entitled to a judicial review of this determination in a United States District Court in accordance with 5 U.S.C. § 552.

Sincerely,

Charles Barlow

From: ████ b6 b3

From: ████████ b3 b6

Sent: Thursday, July 23, 2009 3:17 PM

To: ████ b3 b6

Cc:

Subject: RE: PRIORITY — FOIA Case #F09-0063 - APPEAL - Request for records SEARCH & REVIEW — UNCLASSIFIED//~~FOUO~~ b3 b6

Classification: UNCLASSIFIED//~~FOUO~~

classification: UNCLASSIFIED//~~FOUO~~

Hello ████ b6 b3

IMINT has conducted another search for records (including retired records) regarding this FOIA and IMINT has no records pertaining to "MRI" and "magnetic resonance imaging".

Please let me know if any questions or concerns.

Thanks ████ b6 b3

████████ b6 b3
DNI//████ b3
Room#████ b3

They invented MRI?

From: ████████ b6 b3

Sent: Wednesday, July 22, 2009 12:53 PM

To: ████ b6 b3

Cc: ████████████ FOIA (IART) b6 b3

Subject: PRIORITY — FOIA Case #F09-0063 - APPEAL - Request for records SEARCH & REVIEW — UNCLASSIFIED//~~FOUO~~

Importance: High

classification: UNCLASSIFIED//~~FOUO~~

████ b6 b3 this one is a PRIORITY request.......

9/16/2009

Sent: Tuesday, June 30, 2009 11:07 AM
To: b6 b3
Cc: b6 b3
Subject: RE: FOIA Case #F09-0063 – Request for records SEARCH & REVIEW -- UNCLASSIFIED//FOUO

classification: UNCLASSIFIED//FOUO

Good Morning b6 b3

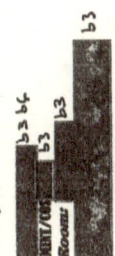

IMINT has no records pertaining to this FOIA case.

Signed case worksheet is attached.

Thank you,

b3 b6
IMINT/CG: b3
Room: b3

From: b3 b6
Sent: Tuesday, June 16, 2009 10:10 AM
To: b3 b6
Cc: FOIA (IART)
Subject: FOIA Case #F09-0063 – Request for records SEARCH & REVIEW -- UNCLASSIFIED//FOUO
Importance: High

classification: UNCLASSIFIED//FOUO

Good morning!

(U) The IART is in receipt of a request for records for:

"1. ...information on functional magnetic resonance imaging.
 2. The date it was put into service.
 3. The first successful report on the first person it was used on successfully."

(U//FOUO) The IART requests that you perform a search for potential responsive documents, and review any documents located for your element's

9/16/2009

Classify Appropriately When Filed In

CASE WORKSHEET

I. CASE INFORMATION

1. CASE NUMBER: P09-0063 APPEAL	2. CASE OFFICER: ▓▓▓ b3 b6	3. SUSPENSE DATE: 30 Jul 2009

4. TYPE OF REVIEW:

- ☒ FREEDOM OF INFORMATION ACT
- ☐ PRIVACY ACT
- ☐ MANDATORY DECLASSIFICATION

5. TYPE OF REQUEST:

- ☐ INITIAL
- ☒ APPEAL
- ☐ COORDINATION
- ☐ REFERRAL

6. TASK THE FOLLOWING OFFICES:

☒ AS&T	☐ COO	☒ IMINT	☐ DS&CI/Counterintelligence	☒ SIGINT
☐ BPO	☒ DDMS	☐ NS&O	☐ OS&CI/Policy	☐ SO
☐ BPO/Contracts	☐ EEO	☐ MS&O/ART	☐ OS&CI/PSD	☐ OTHER: _____
☐ CIO	☐ GED	☐ OOC	☐ OSL	
☐ COMM	☐ IG	☐ OSC/OCC	☒ SE	

7. FEE CATEGORY: ☐ COMMERCIAL ☐ NEWS MEDIA/EDUCATIONAL/SCIENTIFIC ☒ OTHER

Please review for release any document(s) found in response to a Freedom of Information Act request. Mark either your concurrence with the proposed release or non-concurrence with reasoning or justification. Reasoning or justification must be supported by one of the nine FOIA exemptions. See Exemptions page.

II. COMPONENT RESPONSE

1. RECORDS FOUND: ☐ YES ☒ NO

	SEARCH HOURS	REVIEW HOURS		SEARCH METHOD
2. IS 2-4; GS 05-08; E1-E9	11			☒ ELECTRONIC DATABASES
GS/M 09-15; 01-06				☒ LIBRARIES/ARCHIVES (FILE PLAN)
GS/M 16 & Above; 07-010				☒ TASKED TECHNICAL EXPERTS
				☒ SURVEYED ALL EMPLOYEES
				☐ OTHER: (Specify) _____

3. ACTION RECOMMENDED: ☐ GRANT IN FULL ☐ GRANT IN PART ☐ DENY ☐ GLOMAR

4. REMARKS/COMMENTS: *(If additional space is needed, attach separate page.)*

IV. IAO APPROVAL

1. NAME: ▓▓▓▓ b3 b6	
(Type or print full name)	
2. DATE: 27 Jul 2009	3. SIGNATURE: ▓▓▓▓ b3 b6

DECL ON:
DRV FROM:

OPR: MS&O/ASG/MSC/ART

NP11-31, DEC 08 (DS-EF)

Classify Appropriately When Filed In

PREVIOUS EDITIONS ARE OBSOLETE

RCS: Refer to Item 205

b3
b6

From:
Sent: Tuesday, July 28, 2009 9:39 AM
To: FOIA (IART)
Cc:
Subject: RE: PRIORITY -- FOIA Case #F09-0063 -APPEAL -- UNCLASSIFIED//FOUO
Attachments: F09-0063.xfdl
Classification: UNCLASSIFIED//FOUO

classification: UNCLASSIFIED//FOUO

No records found in DDMs relating to "magnetic resonance imaging"

b3
b6

Thank you,

b3
b6

ED/MS/CoS

From:
Sent: Wednesday, July 22, 2009 1:05 PM
To: SE-DAG-INBOX; ast-dag; FOIA (IART)
Cc:
Subject: PRIORITY -- FOIA Case #F09-0063 -APPEAL -- UNCLASSIFIED//FOUO
Importance: High

b3
b6

classification: UNCLASSIFIED//FOUO

(U//FOUO) The IART has received an APPEAL to a "no records" response to a 14 May, 2009 request for "...information on functional magnetic resonance imaging...."

(U//FOUO) The IART has accepted the appeal. We request that each tasked element search for *any records that may exist in their holdings pertaining to magnetic resonance imaging.* The search should include hard copy and soft copy records and documents, electronic databases,

9/16/2009

From: [redacted] b3 b6

Sent: Tuesday, September 01, 2009 3:29 PM

To: [redacted] b3 b6

Cc: FOIA (IART)

Subject: RE: PRIORITY – FOIA Case #F09-0063 -APPEAL (Gilbert Roman) — SECRET//TALENT KEYHOLE//25X1 20590901 RRG dated July 2005

Classification: SECRET//TALENT KEYHOLE//25X1 20590901 RRG dated July 2005

classification: SECRET//TALENT KEYHOLE//25X1 20590901 RRG dated July 2005

[redacted] b3 b6

(//FOUO) Having conducted a search to the best of our abilities across several Ds & Os, we think we've at least met the threshold for the definition of a "reasonable search." We're going to move ahead recommending that the NRO appeal authority uphold our initial "no records" response.... which means you can close your action on this case.

Thanks for your help!

[redacted] b3 b6

[redacted] b3 b6
Senior Case Analyst
IHS/C/IART
[redacted] b3

From: [redacted] b3 b6

Sent: Tuesday, September 01, 2009 2:42 PM

To: [redacted] b3 b6

Subject: FW: PRIORITY – FOIA Case #F09-0063 -APPEAL (Gilbert Roman) — SECRET//TALENT KEYHOLE//25X1 20590901 RRG dated July 2005

classification: SECRET//TALENT KEYHOLE//25X1 20590901 RRG dated July 2005

[redacted] b3 b6

Here's some more info that one of our guys has uncovered. This project predates AS&T, so I don't know how much we'll be able to dig up.

Thanks,

b3
b6

From:
Sent: Tuesday, September 01, 2009 2:26 PM
To:
Subject: FW: PRIORITY -- FOIA Case #F09-0063 -APPEAL (Gilbert Roman) -- SECRET//TALENT KEYHOLE//25X1 20590901 RRG dated July 2005

classification: SECRET//TALENT KEYHOLE//25X1 20590901 RRG dated July 2005

(U) More information on the MRI.

From:
Sent: Tuesday, September 01, 2009 2:25 PM
To:
Subject: FW: PRIORITY -- FOIA Case #F09-0063 -APPEAL (Gilbert Roman) -- SECRET//TALENT KEYHOLE//25X1 20590901 RRG dated July 2005

classification: SECRET//TALENT KEYHOLE//25X1 20590901 RRG dated July 2005

(S//TK) Superconductivity Project No. was RT 23C009. RT=Reconnaissance Technology; 23 = Electronics; C=Program C (Navy); 009=an assigned project number.

Files act hidden in a two step form

From:
Sent: Tuesday, September 01, 2009 2:15 PM
To:
Subject: RE: PRIORITY -- FOIA Case #F09-0063 -APPEAL (Gilbert Roman) -- SECRET//TALENT KEYHOLE//25X1 20590901 RRG dated July 2005

classification: SECRET//TALENT KEYHOLE//25X1 20590901 RRG dated July 2005

9/16/2009

b6

b)(3)

From: b6 b)(3)
Sent: Tuesday, June 30, 2009 11:07 AM
To: b6 b)(3)
Cc: b6 b)(3)
Subject: RE: FOIA Case #F09-0063 - Request for records SEARCH & REVIEW — UNCLASSIFIED//FOUO
Attachments: case Worksheet.xdd
Classification: UNCLASSIFIED//FOUO

classification: UNCLASSIFIED//FOUO

Good Morning b)(3) b6

IMINT has no records pertaining to this FOIA case.

Signed case worksheet is attached.

Thank you,

b)(3) b6
b)(3)
Room: b)(3)
b)(3)

---

b6
From: b)(3)
Sent: Tuesday, June 16, 2009 10:10 AM
To: b6 b)(3)
Cc: b6 b)(3) FOIA (IART)
Subject: FOIA Case #F09-0063 - Request for records SEARCH & REVIEW — UNCLASSIFIED//FOUO
Importance: High

classification: UNCLASSIFIED//FOUO

Good morning!

(U) The IART is in receipt of a request for records for:

"1. ...information on functional magnetic resonance imaging.
2. The date it was put into service.
3. The first successful report on the first person it was used on successfully."

(U//FOUO) The IART requests that you perform a search for potential responsive documents, and review any documents located for your element's equities.
Specifically, we need line-by-line review of any responsive documents for information about your element's equities that should be withheld under applicable FOIA exemptions.

(U) The Case Worksheet provides an explanation of the FOIA exemptions. While we do not require reviewers to

6/30/2009

cite specific FOIA exemptions, we do need you to provide a brief explanation of the rationale for _each individual instance_ in which information is being withheld. Remember that it is not always legitimate to withhold entire paragraphs simply because of their portion markings. The FOIA, and the presumption of disclosure as directed by Executive Order require that we release reasonably segregable portions of text even within classified or otherwise-sensitive paragraphs. Please contact me if you have questions about proper redaction techniques.

(U) Please be sure to provide the IART with information on search methods, and figures for all search and review time expended in processing this request (see Case Worksheet.) In the event that your search time reaches 2 hours, please contact me before expending any additional time.

(U//FOUO) Please be aware that your element's review is only part of the initial phase of the release process. The IART analyzes all responsive documents to determine which elements/agencies should be included in the review process. We welcome your suggestions in this regard, but we ask that you do not task review outside your element.

(U) Should you, or anyone involved in this review, have any questions, please feel free to contact me. **We respectfully request your response NLT 30 June 2009.**

(U) Attachments:
1. Request
2. Case Worksheet

Thanks.
b 6

(b)(3)

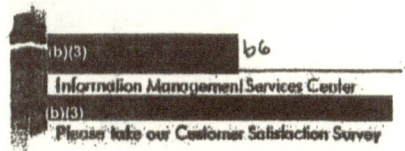
(b)(3)   b6
Information Management Services Center
(b)(3)
Please take our Customer Satisfaction Survey

UNCLASSIFIED//FOUO

UNCLASSIFIED//FOUO

6/30/2009

1. CASE NUMBER: F09-0063   **2. CASE OFFICER:** (b)(3)   **3. SUSPENSE DATE:** 30 Jun 2009

4. TYPE OF REVIEW:

- [X] FREEDOM OF INFORMATION ACT
- [ ] PRIVACY ACT
- [ ] MANDATORY DECLASSIFICATION

5. TYPE OF REQUEST:

- [X] INITIAL
- [ ] APPEAL
- [ ] COORDINATION
- [ ] REFERRAL

6. TASK THE FOLLOWING OFFICES:

[ ] AS&T	[ ] COO	[X] IMINT	[ ] OS&CI/Counterintelligence	[ ] SIGINT
[ ] BPO	[ ] DDMS	[ ] MS&O	[ ] OS&CI/Policy	[ ] SO
[ ] BPO/Contracts	[ ] EEO	[ ] MS&O/ART	[ ] OS&CI/PSD	[ ] OTHER: _____
[ ] CIO	[ ] GED	[ ] OGC	[ ] OSL	
[ ] COMM	[ ] IG	[ ] OSC/OCC	[ ] SE	

7. FEE CATEGORY: [ ] COMMERCIAL   [ ] NEWS MEDIA/EDUCATIONAL/SCIENTIFIC   [X] OTHER

Please review for release any document(s) found in response to a Freedom of Information Act request. Mark either your concurrence with the proposed release or non-concurrence with reasoning or justification. Reasoning or justification must be supported by one of the nine FOIA exemptions. See Exemptions page.

1. RECORDS FOUND: [ ] YES   [ ] NO

2.

	SEARCH HOURS	REVIEW HOURS	SEARCH METHOD
IS 2-4; GS 03-06; E1-E9			[ ] ELECTRONIC DATABASES
GS/M 09-15; 01-06			[ ] LIBRARIES/ARCHIVES (FILE PLAN)
GS/M 16 & Above; 07-010			[ ] TASKED TECHNICAL EXPERTS
			[ ] SURVEYED ALL EMPLOYEES
			[ ] OTHER: (Specify) _____

3. ACTION RECOMMENDED: [ ] GRANT IN FULL   [ ] GRANT IN PART   [ ] DENY   [ ] GLOMAR

4. REMARKS/COMMENTS: (If additional space is needed, attach separate page.)

NO RECORDS found

1. NAME: (b)(3)

(Type or print full name)

2. DATE: 30 Jun 2009   **3. SIGNATURE:** (b)(3)

L ON:

N FROM:

OPR: MS&O/ASG/IMSC/ART

NP11-31, DEC 08 (DS-EF)

PREVIOUS EDITIONS ARE OBSOLETE

RCS: Refer to Item 205

From: b6 (b)(3)
Sent: Wednesday, June 03, 2009 12:59 PM
To: Moffett Page P NRO USA CIV (b)(3) b6
Cc: FOIA (IART)
Subject: Disgruntled FOIA Requester — UNCLASSIFIED
Attachments: Roman.pdf
Classification: UNCLASSIFIED

classification: UNCLASSIFIED

I am really sorry to bother you with this, Sir, but felt that we should at least give you an FYI . . . so I've attached a complete copy of correspondence we received today via certified mail regarding a response to a FOIA request. The attachment tells the entire story. I'm still debating with myself about how or if we will respond. *My last? paper work.*

I will be looking into why we did not receive the first request that was obviously signed for by the NRO.

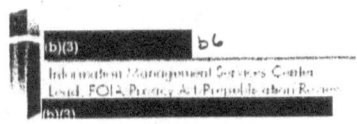

(b)(3) b6
Information Management Services Center
Lead, FOIA, Privacy Act, Prepublication Review
(b)(3)

Please take our Customer Satisfaction Survey

UNCLASSIFIED

6/3/2009

b6
(b)(3)

From:	b6 (b)(3)
Sent:	Wednesday, June 03, 2009 1:10 PM
To:	b6 (b)(3)
Cc:	Hathaway Linda sue hathaway S NRO USA CIV; Jung Steve A NRO USA CIV
Subject:	Certified mail lost — UNCLASSIFIED
Attachments:	signaturecard.pdf
Classification:	UNCLASSIFIED

classification: UNCLASSIFIED

b6

Hi, (b)(3) Need your help, please . . . attached is a copy of receipts that a Freedom of Information Act requester received back from the NRO for certified mail. The first one that was delivered and signed for on 9 March was never received by this office.

The second letter was received today, 3 June, but appears to have been signed for on 8 May even though the postmark on the envelope is 27 May and the date on the requester's letter is 27 May.

The clock on legal time limitations for us to respond starts ticking when the letter is received at the NRO. We have 20 days to respond and a lot work to do in that time. Any delay for us could be costly. This requester is now threatening to sue because we didn't respond.

Can you please see if you can determine why we never received the first letter and why it appears to have taken a week for it to (b)(3) b6

Thank you so much! I know there's nothing we can do to fix this particular problem, but in the future, if necessary, we can (b)(3) and the time issue is critical for us from a legal perspective.

b6

(b)(3)
Information Management Services Center
Lead, FOIA/Privacy Act/Prepublication Review
(b)(3)

Please take our Customer Satisfaction Survey

UNCLASSIFIED

6/3/2009

b6

(b)(3)

From:	b6 (b)(3)
Sent:	Tuesday, June 30, 2009 11:07 AM
To:	b6 (b)(3)
Cc:	b6 (b)(3)
Subject:	RE: FOIA Case #F09-0063 - Request for records SEARCH & REVIEW --- UNCLASSIFIED//FOUO
Attachments:	case Worksheet.xfdl
Classification:	UNCLASSIFIED//FOUO

classification: UNCLASSIFIED//FOUO

Good Morning (b)(3) b6

IMINT has no records pertaining to this FOIA case.

Signed case worksheet is attached.

Thank you,
b6

(b)(3)
IMINT/COS (b)(3)
Room: (b)(3)
(b)(3)

---

b6

From: (b)(3)
Sent: Tuesday, June 16, 2009 10:10 AM
To: (b)(3)   b6
b6 **Cc:** (b)(3)   FOIA (IART)
Subject: FOIA Case #F09-0063 - Request for records SEARCH & REVIEW --- UNCLASSIFIED//FOUO
Importance: High

classification: UNCLASSIFIED//FOUO

Good morning!

(U) The IART is in receipt of a request for records for:

> "1. ...information on functional magnetic resonance imaging.
> 2. The date it was put into service.
> 3. The first successful report on the first person it was used on successfully."

(U//FOUO) The IART requests that you perform a search for potential responsive documents, and review any documents located for your element's equities.
Specifically, we need line-by-line review of any responsive documents for information about your element's equities that should be withheld under applicable FOIA exemptions.

(U) The Case Worksheet provides an explanation of the FOIA exemptions. While we do not require reviewers to

6/30/2009

cite specific FOIA exemptions, we do need you to provide a brief explanation of the rationale for _each individual instance_ in which information is being withheld. Remember that it is not always legitimate to withhold entire paragraphs simply because of their portion markings. The FOIA, and the presumption of disclosure as directed by Executive Order require that we release reasonably segregable portions of text even within classified or otherwise-sensitive paragraphs. Please contact me if you have questions about proper redaction techniques.

(U) Please be sure to provide the IART with information on search methods, and figures for all search and review time expended in processing this request (see Case Worksheet.) In the event that your search time reaches 2 hours, please contact me before expending any additional time.

(U//FOUO) Please be aware that your element's review is only part of the initial phase of the release process. The IART analyzes all responsive documents to determine which elements/agencies should be included in the review process. We welcome your suggestions in this regard, but we ask that you do not task review outside your element.

(U) Should you, or anyone involved in this review, have any questions, please feel free to contact me. **We respectfully request your response NLT 30 June 2009.**

(U) Attachments:
1. Request
2. Case Worksheet

Thanks.

 b6

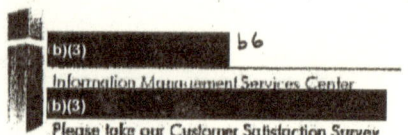
b6
Information Management Services Center
Please take our Customer Satisfaction Survey

UNCLASSIFIED//FOUO

UNCLASSIFIED//FOUO

6/30/2009

CASE WORKSHEET

I. CASE INFORMATION

1. CASE NUMBER: F09-0063   2. CASE OFFICER: (b)(3) b6   3. SUSPENSE DATE: 30 Jun 2009

4. TYPE OF REVIEW:

- ☒ FREEDOM OF INFORMATION ACT
- ☐ PRIVACY ACT
- ☐ MANDATORY DECLASSIFICATION

5. TYPE OF REQUEST:

- ☒ INITIAL
- ☐ APPEAL
- ☐ COORDINATION
- ☐ REFERRAL

6. TASK THE FOLLOWING OFFICES:

- ☐ AS&T
- ☐ BPO
- ☐ BPO/Contracts
- ☐ CIO
- ☐ COMM
- ☐ COO
- ☐ DDMS
- ☐ EEO
- ☐ GED
- ☐ IG
- ☒ IMINT
- ☐ MS&O
- ☐ MS&O/IART
- ☐ OGC
- ☐ OSC/OOO
- ☐ OS&CI/Counterintelligence
- ☐ OS&CI/Policy
- ☐ OS&CI/PSD
- ☐ OSL
- ☐ SE
- ☐ SIGINT
- ☐ SO
- ☐ OTHER: _____

7. FEE CATEGORY: ☐ COMMERCIAL   ☐ NEWS MEDIA/EDUCATIONAL/SCIENTIFIC   ☒ OTHER

Please review for release any document(s) found in response to a Freedom of Information Act request. Mark either your concurrence with the proposed release or non-concurrence with reasoning or justification. Reasoning or justification must be supported by one of the nine FOIA exemptions. See Exemptions page.

II. COMPONENT RESPONSE

1. RECORDS FOUND: ☐ YES   ☐ NO

	SEARCH HOURS	REVIEW HOURS	SEARCH METHOD
2. IS 2-4; GS 03-08; E1-E9			☐ ELECTRONIC DATABASES
GS/M 09-15; 01-06	5		☐ LIBRARIES/ARCHIVES (FILE PLAN)
GS/M 16 & Above; 07-010			☐ TASKED TECHNICAL EXPERTS
			☐ SURVEYED ALL EMPLOYEES
			☐ OTHER: (Specify) _____

3. ACTION RECOMMENDED: ☐ GRANT IN FULL   ☐ GRANT IN PART   ☐ DENY   ☐ GLOMAR

4. REMARKS/COMMENTS: *(If additional space is needed, attach separate page.)*

NO RECORDS found

III. IAD APPROVAL

1. NAME: (b)(3) b6
   (Type or print full name)   b6

2. DATE: 30 Jun 2009   3. SIGNATURE: (b)(3)

DECL ON:

DRV FROM:

OPR: MS&O/ASG/IMSC/IART

NP11-31, DEC 08 (DS-EF)

CASE WORKSHEET (Continued)

EXPLANATION OF EXEMPTIONS

The designation "(b)(1)" applies to information which is currently and properly classified pursuant to an Executive Order in the interest of national defense or foreign policy (EO 12958 and DoD Reg 5200.1-R).

Sec 1.4 Classification Categories:

Information may NOT be considered for classification unless it concerns:
a) military plans, weapons systems, or operations;
b) foreign government information;
c) intelligence activities (including special activities), intelligence sources or methods, or cryptology;
d) foreign relations or foreign activities of the United States including confidential sources;
e) scientific, technological, or economic matters relating to the national security;
f) United States Government programs for safeguarding nuclear materials or facilities;
g) vulnerabilities or capabilities of systems, installations, projects or plans relating to national security;
h) weapons of mass destruction.

The designation "(b)(2)" applies to information which pertains solely to the internal rules and practices of an agency; this exemption has two profiles, "high" and "low". The "high" profile permits withholding of a document which, if released, would allow circumvention of an organization rule, policy, or statute, thereby impeding the agency in the conduct of its mission. The "low" profile permits withholding if there is no public interest in the document, and it would be an administrative burden to process the request.

The designation "(b)(3)" applies to information specifically exempted by a statute establishing a particular criteria for withholding. The language of the statute must clearly state that the information will not be disclosed.

The designation "(b)(4)" applies to information such as trade secrets and commercial or financial information obtained from a company on a privileged or confidential basis which, if released, would result in competitive harm to the company.

The designation "(b)(5)" applies to inter-agency and intra-agency memoranda which are deliberative in nature; this exemption is appropriate for internal documents which are part of the decisionmaking process and contain subjective evaluations, opinions, and recommendations.

The designation "(b)(6)" applies to information release of which could reasonably be expected to constitute a clearly unwarranted invasion of the personal privacy of individuals.

The designation "(b)(7)" applies to records or information compiled for law enforcement purposes and that (a) could reasonably be expected to interfere with law enforcement proceedings; (b) would deprive a person of a right to a fair trial or impartial adjudication; (c) could reasonably be expected to constitute an unwarranted invasion of the personal privacy of others; (d) could disclose the identity of a confidential source; (e) could disclose investigative techniques and procedures; or (f) could reasonably be expected to endanger the life or physical safety of an individual.

The designation "(b)(8)" applies to matters that are "contained in or related to examination, operating, or condition reports prepared by on behalf of or for use of any agency responsible for the regulation of supervision of financial institutions. (Not used by the NRO.)

The designation "(b)(9)" applies to geological information and data, including maps, concerning wells. (Not used by the NRO.)

b6
(b)(3)

From:	b6 (b)(3)
Sent:	Tuesday, June 16, 2009 10:10 AM
To:	b6 (b)(3)
Cc:	b6 (b)(3) FOIA (IART)
Subject:	FOIA Case #F09-0063 - Request for records SEARCH & REVIEW --- UNCLASSIFIED//FOUO

Importance: High

Attachments: F09-0063 Request.tif; F09-0063_Case_Worksheet.xfdl

Classification: UNCLASSIFIED//FOUO

classification: UNCLASSIFIED//FOUO

Good morning!

(U) The IART is in receipt of a request for records for:

> "1. ...information on functional magnetic resonance imaging.
> 2. The date it was put into service.
> 3. The first successful report on the first person it was used
> on successfully."

(U//FOUO) The IART requests that you perform a search for potential responsive documents, and review any documents located for your element's equities.
Specifically, we need line-by-line review of any responsive documents for information about your element's equities that should be withheld under applicable FOIA exemptions.

(U) The Case Worksheet provides an explanation of the FOIA exemptions. While we do not require reviewers to cite specific FOIA exemptions, we do need you to provide a brief explanation of the rationale for *each individual instance* in which information is being withheld. Remember that it is not always legitimate to withhold entire paragraphs simply because of their portion markings. The FOIA, and the presumption of disclosure as directed by Executive Order require that we release reasonably segregable portions of text even within classified or otherwise-sensitive paragraphs. Please contact me if you have questions about proper redaction techniques.

(U) Please be sure to provide the IART with information on search methods, and figures for all search and review time expended in processing this request (see Case Worksheet.) In the event that your search time reaches 2 hours, please contact me before expending any additional time.

(U//FOUO) Please be aware that your element's review is only part of the initial phase of the release process. The IART analyzes all responsive documents to determine which elements/agencies should be included in the review process. We welcome your suggestions in this regard, but we ask that you do not task review outside your element.

(U) Should you, or anyone involved in this review, have any questions, please feel free to contact me. **We respectfully request your response NLT 30 June 2009.**

(U) Attachments:
1. Request
2. Case Worksheet

6/16/2009

b6
(b)(3)

From:	b6 (b)(3)
Sent:	Thursday, October 15, 2009 11:51 AM
To:	b6 (b)(3)
Cc:	FOIA (IART)
Subject:	F09-0063 Appeal/litigation & F-09-0064— UNCLASSIFIED//FOUO — UNCLASSIFIED//FOUO

Classification: UNCLASSIFIED//FOUO

classification: UNCLASSIFIED//FOUO

would not sign without changes

Dave,

Mr. Barlow signed the appeal letter for F09-063 this morning.

Mr. Barlow **did not** sign the F09-0064 appeal letter. He asked the IG (b)(3) to reconsider redacting portions of the responsive documents for release to the requestor. (b)(3) was going to work on it and get back to us ASAP. I will keep you apprised on this case.

F09-0063 is ready for pick-up.
F09-0074 is ready for pick-up (b)(3) case).
Also providing you a copy of attny appeal letter re: F09-0064 for your files.

b6
(b)(3)
Office of General Counsel
(b)(3)

From: (b)(3)
Sent: Tuesday, October 06, 2009 8:35 AM
To: (b)(3) b6
Cc: FOIA (IART)
Subject: RE: Gilbert Roman and F09-0063 Appeal/litigation --- UNCLASSIFIED//FOUO

classification: UNCLASSIFIED//FOUO

Good morning, (b)(3)
b6

(U) The Appeal response letters for F09-0063 and F09-0064 are attached. Let me know if you need anything additional..... thanks for all your help!

(b)(3)
b6

(b)(3) b6
Senior Case Analyst
IMSC/IART

10/15/2009

b6
(b)(3)

---

b6
From: (b)(3)
Sent: Tuesday, October 06, 2009 7:55 AM
b6 To: (b)(3)
Cc: FOIA (IART); Moffett Page P NRO USA CIV
Subject: RE: Gilbert Roman and F09-0063 Appeal/litigation --- UNCLASSIFIED//FOUO

classification: UNCLASSIFIED//FOUO

Dave,

FOIA Appeal - F09-0063--GC concurs. Please send me soft copy of the letter for Mr. Barlow's signature and we will take care of from here. FYI-We are meeting w/Mr. Barlow re: Appeal F09-0064 on Tues. 13 Oct and we will take care of this one then also.

Thanks much!

b6
(b)(3)
Office of General Counsel
b6 (b)(3)

---

b6
From: (b)(3)
Sent: Tuesday, September 01, 2009 6:51 AM
To: (b)(3) b6
Cc: FOIA (IART)
Subject: Gilbert Roman and F09-0063 Appeal/litigation --- UNCLASSIFIED//FOUO

classification: UNCLASSIFIED//FOUO
b6
(U//FOUO) Good morning (b)(3) This is just a brief update on where we stand on the processing of Mr. Gilbert's appeal of our "no records" response in F09-0063. I'm still waiting for a search/review response from a couple of the tasked Ds/Os. So far, IMINT, DDMS, and SIGINT have located no documents responsive to his request for "information on functional magnetic resonance imaging." I'm trying to corral the remaining responses so I can move a final package forward.
b6
(b)(3)
b6
(b)(3)
Senior Case Analyst
IMSC/IART
(b)(3)

UNCLASSIFIED//FOUO

UNCLASSIFIED//FOUO

10/15/2009

From:	b6 (b)(3)
Sent:	Wednesday, July 22, 2009 12:53 PM
To:	b6 (b)(3)
Cc:	b6 (b)(3) FOIA (IART)
Subject:	PRIORITY ---- FOIA Case #F09-0063 - APPEAL - Request for records SEARCH & REVIEW --- UNCLASSIFIED//FOUO
Importance:	High
Attachments:	F09-0063_APPEAL_Case_Worksheet.xfdl
Classification:	UNCLASSIFIED//FOUO

classification: UNCLASSIFIED//FOUO

b6 (b)(3) this one is a PRIORITY request......

(U//FOUO) The requester in FOIA case #F09-0063 has appealed the NRO's "no records" response to his 14 May, 2009 request for "...Information on functional magnetic resonance imaging...."

(U//FOUO) The IART has accepted the appeal. We request that IMINT perform a new search for _any_ records that may exist in IMINT holdings pertaining to magnetic resonance imaging. The search should include hard copy and soft copy records and documents, electronic databases, and RETIRED RECORDS. In the event that responsive document are located, we request that IMINT perform an initial review for any sensitive IMINT equities. We will need clean copies of responsive documents as well as copies treated for IMINT equities, if applicable. Please do not task review outside your directorate; your suggestions for additional review are welcome. OS&CI, OSC, and OGC perform a final review of all responsive documents prior to any final release determination.

(U//FOUO) In response to the requester's appeal, the IART is also requesting that DDMS, SIGINT, SE and AS&T search for documents. While we ask that you do not task this request outside your directorate, the IART welcomes your suggestions for other possible sources for responsive documents within the NRO.

(U//FOUO) IMPORTANT INFORMATION ABOUT TREATMENT OF DOCUMENTS FOR RELEASE UNDER THE FOIA:
See the Case Worksheet for the FOIA exemptions that can be invoked to withhold information. The FOIA requires that reasonably segregable releasable information within classified/sensitive documents/portions of documents be released; complete redaction of classified portions of documents is usually not an appropriate treatment. Except for the classified information exemption ((b)(1)), invocation of exemptions is discretionary; if there is no harm in release, then release is appropriate. The IART is required by law to cite FOIA exemptions to justify each individual redaction in a released document. We do not require program office reviewers to cite specific FOIA exemptions, but we do need the rationale for every instance in which you recommend that information be withheld, so the IART can apply the appropriate exemptions. If anyone involved in the review process has questions about treating documents for release, please contact me.

(U//FOUO) Please record details regarding search time/methods on the Case Worksheet provided. So that we may respond to the appeal in a timely manner and mitigate the risk of litigation, we request that you handle this request as a priority, and respond to the IART no later than 30 July 2009.

(U) If you have any questions, please contact me and reference case #F09-0063.

Thanks.

(b)(3) b6

7/22/2009

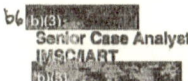

b6 b(3)
Senior Case Analyst
IMSC/IART
b(3)

_____ b6 _____

From: b(3)
Sent: Tuesday, June 30, 2009 11:07 AM
b6 **To:** b(3)
b6 **Cc** b(3)
Subject: RE: FOIA Case #F09-0063 - Request for records SEARCH & REVIEW --- UNCLASSIFIED//FOUO

classification: UNCLASSIFIED//FOUO

Good Morning b(3) b6

IMINT has no records pertaining to this FOIA case.

Signed case worksheet is attached.

Thank you,

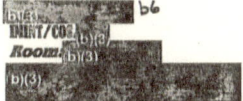

b(3) b6
IMINT/COS b(3)
Room: b(3)
b(3)

_____

From b(3) b6
Sent: Tuesday, June 16, 2009 10:10 AM
To: b(3) b6
b6 **Cc** b(3) , FOIA (IART)
Subject: FOIA Case #F09-0063 - Request for records SEARCH & REVIEW --- UNCLASSIFIED//FOUO
Importance: High

classification: UNCLASSIFIED//FOUO

Good morning!

(U) The IART is in receipt of a request for records for:

        "1. ...information on functional magnetic resonance imaging.
        2. The date it was put into service.
        3. The first successful report on the first person it was used
on successfully."

(U//FOUO) The IART requests that you perform a search for potential responsive documents, and review any documents located for your element's equities.
Specifically, we need line-by-line review of any responsive documents for information about your element's equities that should be withheld under applicable FOIA exemptions.

(U) The Case Worksheet provides an explanation of the FOIA exemptions. While we do not require reviewers to cite specific FOIA exemptions, we do need you to provide a brief explanation of the rationale for each individual instance in which information is being withheld. Remember that it is not always legitimate to withhold entire

7/22/2009

paragraphs simply because of their portion markings. The FOIA, and the presumption of disclosure as directed by Executive Order require that we release reasonably segregable portions of text even within classified or otherwise-sensitive paragraphs. Please contact me if you have questions about proper redaction techniques.

(U) Please be sure to provide the IART with information on search methods, and figures for all search and review time expended in processing this request (see Case Worksheet.) In the event that your search time reaches 2 hours, please contact me before expending any additional time.

(U//FOUO) Please be aware that your element's review is only part of the initial phase of the release process. The IART analyzes all responsive documents to determine which elements/agencies should be included in the review process. We welcome your suggestions in this regard, but we ask that you do not task review outside your element.

(U) Should you, or anyone involved in this review, have any questions, please feel free to contact me. **We respectfully request your response NLT 30 June 2009.**

(U) Attachments:
1. Request
2. Case Worksheet

Thanks.

(b)(3) b6

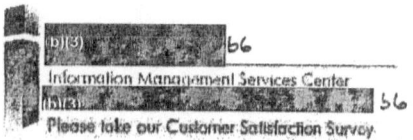

(b)(3) b6
Information Management Services Center
(b)(3) b6
Please take our Customer Satisfaction Survey.

UNCLASSIFIED//FOUO

UNCLASSIFIED//FOUO

UNCLASSIFIED//FOUO

7/22/2009

b6
(b)(3)

_____

From: b6 (b)(3)
Sent: Wednesday, July 22, 2009 1:05 PM
To: b6 (b)(3)
Cc: b6 (b)(3) SE-DAG-INBOX; ast-dag; FOIA (IART)
Subject: PRIORITY -- FOIA Case #F09-0063 -APPEAL --- UNCLASSIFIED//FOUO
Importance: High
Attachments: F09-0063_APPEAL_Case_Worksheet.xldl
Classification: UNCLASSIFIED//FOUO

classification: UNCLASSIFIED//FOUO

(U//FOUO) The IART has received an APPEAL to a "no records" response to a 14 May, 2009 request for "...**information on functional magnetic resonance imaging....**"

(U//FOUO) The IART has accepted the appeal. **We request that each tasked element search for _any records that may exist in their holdings pertaining to magnetic resonance imaging._ The search should include hard copy and soft copy records and documents, electronic databases, and RETIRED RECORDS.** In the event that responsive document are located, we request that each element perform an initial review for their own equities that should be withheld from release. We will need clean copies of responsive documents and copies treated for sensitive equities, if applicable. Please do not task review outside your directorate; your suggestions for additional element review(s) are welcome. OS&CI, OSC, and OGC perform a final review of all responsive documents prior to any final release determination.

(U//FOUO) In the initial processing of this request, IMINT was tasked to search for responsive documents. In response to the requester's appeal, **IMINT** is performing a new search, and the IART is requesting that **DDMS, SIGINT, SE and AS&T** also search for documents. While we ask that you do not task this search outside your directorate, the IART welcomes your suggestions for other possible sources for responsive documents within the NRO.

(U//FOUO) **IMPORTANT INFORMATION ABOUT TREATMENT OF DOCUMENTS FOR RELEASE UNDER THE FOIA:**
See the Case Worksheet for the FOIA exemptions that can be invoked to withhold information. The FOIA requires that reasonably segregable releasable information within classified/sensitive documents/portions of documents be released; complete redaction of classified portions of documents is usually not an appropriate treatment. Except for the classified information exemption ((b)(1)), invocation of exemptions is discretionary; if there is no harm in release, then release is appropriate. The IART is required by law to cite FOIA exemptions to justify each individual redaction in a released document. We do not require program office reviewers to cite specific FOIA exemptions, but we do need the rationale for every instance in which you recommend that information be withheld, so the IART can apply the appropriate exemptions. If anyone involved in the review process has questions about treating documents for release, please contact me.

(U//FOUO) Please record details regarding search time/methods on the Case Worksheet provided. **So that we may respond to the appeal in a timely manner and mitigate the risk of litigation, we request that you handle this request as a priority, and respond to the IART no later than 30 July 2009.**

(U) If you have any questions, please contact me and reference case #F09-0063(A).

(b)(3) b6

(b)(3) b6
Senior Case Analyst

7/22/2009

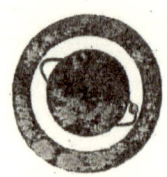

NATIONAL RECONNAISSANCE OFFICE
14675 Lee Road
Chantilly, VA 20151-1715

22 July 2009

Gilbert Roman
95-25 77th Street
Ozone Park, NY 11416

Dear Mr. Roman:

    This is in response to your letter dated 12 July 20098, received in the Information Management Services Center of the National Reconnaissance Office (NRO) on 21 July 2009. Pursuant to the Freedom of Information Act (FOIA), you are appealing the NRO's response to your request for "...information on functional magnetic resonance imaging..."

    Your appeal has been accepted. We will advise you when a determination by the NRO Appeal Board has been made.

    If you have any questions, please call the Requester Service Center at (703) 227-9326, and reference case number F09-0063.

                                    Sincerely,

                                    Linda S. Hathaway
                                    Chief, Information Access
                                    and Release Team

From: b6 (b)(3)
Sent: Thursday, July 23, 2009 3:17 PM
To: b6 (b)(3)
Cc: b6 (b)(3)

Subject: RE: PRIORITY ---- FOIA Case #F09-0063 - APPEAL - Request for records SEARCH & REVIEW --- UNCLASSIFIED//FOUO

Classification: UNCLASSIFIED//FOUO

classification: UNCLASSIFIED//FOUO

Hello (b)(3) b6

IMINT has conducted another search for records (including retired records) regarding this FOIA and IMINT has <u>no</u> records pertaining to "MRI" and "magnetic resonance imaging" .

Please let me know if any questions or concerns.

Thanks (b)(3) b6

(b)(3) b6
IMINT/COS
Room: (b)(3)
(b)(3)

From: (b)(3) b6
Sent: Wednesday, July 22, 2009 12:53 PM
To: (b)(3) b6
b6 Cc: (b)(3) FOIA (IART)
Subject: PRIORITY ---- FOIA Case #F09-0063 - APPEAL - Request for records SEARCH & REVIEW --- UNCLASSIFIED//FOUO
Importance: High

classification: UNCLASSIFIED//FOUO

b6 (b)(3) this one is a PRIORITY request......

(U//FOUO) The requester in FOIA case #F09-0063 has appealed the NRO's "no records" response to his 14 May, 2009 request for "...**information on functional magnetic resonance imaging....**"

(U//FOUO) The IART has accepted the appeal. **We request that IMINT perform a new search for <u>any records</u> that may exist in IMINT holdings pertaining to magnetic resonance imaging. The search should include hard copy and soft copy records and documents, electronic databases, and RETIRED RECORDS.** In the event that responsive document are located, we request that IMINT perform an intitial review for any sensitive IMINT equities. We will need clean copies of responsive documents as well as copies treated for IMINT equities, if applicable. Please do not task review outside your directorate; your suggestions for additional review are welcome. OS&CI, OSC, and OGC perform a final review of all responsive documents prior to any final release determination.

7/24/2009

(U//FOUO) In response to the requester's appeal, the IART is also requesting that DDMS, SIGINT, SE and AS&T search for documents. While we ask that you do not task this request outside your directorate, the IART welcomes your suggestions for other possible sources for responsive documents within the NRO.

(U//FOUO) **IMPORTANT INFORMATION ABOUT TREATMENT OF DOCUMENTS FOR RELEASE UNDER THE FOIA:**
See the Case Worksheet for the FOIA exemptions that can be invoked to withhold information. The FOIA requires that reasonably segregable releasable information within classified/sensitive documents/portions of documents be released; complete redaction of classified portions of documents is usually not an appropriate treatment. Except for the classified information exemption ((b)(1)), invocation of exemptions is discretionary; if there is no harm in release, then release is appropriate. The IART is required by law to cite FOIA exemptions to justify each individual redaction in a released document. We do not require program office reviewers to cite specific FOIA exemptions, but we do need the rationale for every instance in which you recommend that information be withheld, so the IART can apply the appropriate exemptions. If anyone involved in the review process has questions about treating documents for release, please contact me.

(U//FOUO) Please record details regarding search time/methods on the Case Worksheet provided. **So that we may respond to the appeal in a timely manner and mitigate the risk of litigation, we request that you handle this request as a priority, and respond to the IART no later than 30 July 2009.**

(U) If you have any questions, please contact me and reference case #F09-0063.

Thanks.

Senior Case Analyst
IMSC/IART

From: (b)(3)
Sent: Tuesday, June 30, 2009 11:07 AM
To: (b)(3)
Cc: (b)(3)
Subject: RE: FOIA Case #F09-0063 - Request for records SEARCH & REVIEW --- UNCLASSIFIED//FOUO

classification: UNCLASSIFIED//FOUO

Good Morning (b)(3)

IMINT has no records pertaining to this FOIA case.

Signed case worksheet is attached.

Thank you,

From: (b)(3)

7/24/2009

Sent: Tuesday, June 16, 2009 10:10 AM
To:
Cc: ████████████████████████████████████ FOIA (IART)
Subject: FOIA Case #F09-0063 - Request for records SEARCH & REVIEW — UNCLASSIFIED//FOUO
Importance: High

classification: UNCLASSIFIED//FOUO

Good morning!

(U) The IART is in receipt of a request for records for:

"1. ...information on functional magnetic resonance imaging.
2. The date it was put into service.
3. The first successful report on the first person it was used
on successfully."

(U//FOUO) The IART requests that you perform a search for potential responsive documents, and review any documents located for your element's equities.
Specifically, we need line-by-line review of any responsive documents for information about your element's equities that should be withheld under applicable FOIA exemptions.

(U) The Case Worksheet provides an explanation of the FOIA exemptions. While we do not require reviewers to cite specific FOIA exemptions, we do need you to provide a brief explanation of the rationale for each individual instance in which information is being withheld. Remember that it is not always legitimate to withhold entire paragraphs simply because of their portion markings. The FOIA, and the presumption of disclosure as directed by Executive Order require that we release reasonably segregable portions of text even within classified or otherwise-sensitive paragraphs. Please contact me if you have questions about proper redaction techniques.

(U) Please be sure to provide the IART with information on search methods, and figures for all search and review time expended in processing this request (see Case Worksheet.) In the event that your search time reaches 2 hours, please contact me before expending any additional time.

(U//FOUO) Please be aware that your element's review is only part of the initial phase of the release process. The IART analyzes all responsive documents to determine which elements/agencies should be included in the review process. We welcome your suggestions in this regard, but we ask that you do not task review outside your element.

(U) Should you, or anyone involved in this review, have any questions, please feel free to contact me. We respectfully request your response NLT 30 June 2009.

(U) Attachments:
1. Request
2. Case Worksheet

Thanks.

Information Management Services Center
Please take our Customer Satisfaction Survey

UNCLASSIFIED//FOUO
UNCLASSIFIED//FOUO
UNCLASSIFIED//FOUO
UNCLASSIFIED//FOUO

66 (b)(3)

From: 66 (b)(3)
Sent: Monday, July 27, 2009 10:35 AM
To: 66 (b)(3)
Subject: RE: PRIORITY – FOIA Case #F09-0063 -APPEAL — UNCLASSIFIED//FOUO
Attachments: Case Worksheet.pdf
Classification: UNCLASSIFIED//FOUO

classification: UNCLASSIFIED//FOUO

 66

Attached you will find the case worksheet. If you have any questions or concerns please feel free to contact me.

Sincerely,

66 (b)(3)

SIGINT\ODIR\COS\DAG

66

From: (b)(3)
Sent: Wednesday, July 22, 2009 1:05 PM
66 To: (b)(3)
66
66 Cc: (b)(3)   SE-DAG-INBOX; ast-dag; FOIA (IART)
Subject: PRIORITY — FOIA Case #F09-0063 -APPEAL — UNCLASSIFIED//FOUO
Importance: High

classification: UNCLASSIFIED//FOUO

(U//FOUO) The IART has received an APPEAL to a "no records" response to a 14 May, 2009 request for
"...information on functional magnetic resonance imaging...."

(U//FOUO) The IART has accepted the appeal. We request that each tasked element search for *any records
that may exist in their holdings pertaining to magnetic resonance imaging*. The search should
include hard copy and soft copy records and documents, electronic databases, and RETIRED
RECORDS. In the event that responsive document are located, we request that each element perform an intitial
review for their own equities that should be withheld from release. We will need clean copies of responsive
documents and copies treated for sensitive equities, if applicable. Please do not task review outside your
directorate; your suggestions for additional element review(s) are welcome. OS&CI, OSC, and OGC perform a
final review of all responsive documents prior to any final release determination.

(U//FOUO) In the initial processing of this request, IMINT was tasked to search for responsive documents. In
response to the requester's appeal, **IMINT** is performing a new search, and the IART is requesting that **DDMS,
SIGINT, SE and AS&T** also search for documents. While we ask that you do not task this search outside your
directorate, the IART welcomes your suggestions for other possible sources for responsive documents within the
NRO.

(U//FOUO) IMPORTANT INFORMATION ABOUT TREATMENT OF DOCUMENTS FOR RELEASE UNDER
THE FOIA!

8/9/2010

See the Case Worksheet for the FOIA exemptions that can be invoked to withhold information. The FOIA requires that reasonably segregable releasable information within classified/sensitive documents/portions of documents be released; complete redaction of classified portions of documents is usually not an appropriate treatment. Except for the classified information exemption ((b)(1)), invocation of exemptions is discretionary; if there is no harm in release, then release is appropriate. The IART is required by law to cite FOIA exemptions to justify each individual redaction in a released document. We do not require program office reviewers to cite specific FOIA exemptions, but we do need the rationale for every instance in which you recommend that information be withheld, so the IART can apply the appropriate exemptions. If anyone involved in the review process has questions about treating documents for release, please contact me.

(U//FOUO) Please record details regarding search time/methods on the Case Worksheet provided. So that we may respond to the appeal in a timely manner and mitigate the risk of litigation, we request that you handle this request as a priority, and respond to the IART no later than 30 July 2009.

(U) If you have any questions, please contact me and reference case #F09-0063(A).

(b)(3) b6

(b)(3) b6
Senior Case Analyst
IMSC/IART
(b)(3)

UNCLASSIFIED//FOUO

UNCLASSIFIED//FOUO

8/9/2010

CASE WORKSHEET

I. CASE INFORMATION

1. CASE NUMBER: F09-0063 APPEAL   2. CASE OFFICER:  b6   3. SUSPENSE DATE: 30 Jul 2009

4. TYPE OF REVIEW:
- ☒ FREEDOM OF INFORMATION ACT
- ☐ PRIVACY ACT
- ☐ MANDATORY DECLASSIFICATION

5. TYPE OF REQUEST:
- ☐ INITIAL
- ☒ APPEAL
- ☐ COORDINATION
- ☐ REFERRAL

6. TASK THE FOLLOWING OFFICES:

☒ AS&T	☐ COO	☒ R&NT	☐ OS&CI/Counterintelligence	☒ SIGINT
☐ SPO	☒ DDMS	☐ MS&O	☐ OS&CI/Policy	☐ SO
☐ BPO/Contracts	☐ EEO	☐ MSS&ART	☐ OS&CI/PSD	☐ OTHER: _____
☐ CIO	☐ GED	☐ OGC	☐ OSL	
☐ COMM	☐ IG	☐ OSC/OCC	☒ SE	

7. FEE CATEGORY: ☐ COMMERCIAL ☐ NEWS MEDIA/EDUCATIONAL/SCIENTIFIC ☒ OTHER

Please review for release any document(s) found in response to a Freedom of Information Act request. Mark either your concurrence with the proposed release or non-concurrence with reasoning or justification. Reasoning or justification must be supported by one of the nine FOIA exemptions. See Exemptions page.

II. COMPONENT RESPONSE

1. RECORDS FOUND: ☐ YES ☒ NO

	SEARCH HOURS	REVIEW HOURS	SEARCH METHOD
2. IС 3-4; GS 03-06; E1-E9	11	_____	☒ ELECTRONIC DATABASES
GS/M 09-15; 01-06	_____	_____	☒ LIBRARIES/ARCHIVES (FILE PLAN)
GS/M 16 & Above; 07-010	_____	_____	☒ TASKED TECHNICAL EXPERTS
			☒ SURVEYED ALL EMPLOYEES
			☐ OTHER: (Specify) _____

3. ACTION RECOMMENDED: ☐ GRANT IN FULL ☐ GRANT IN PART ☐ DENY ☐ GLOMAR

4. REMARKS/COMMENTS: *(If additional space is needed, attach separate page.)*

III. IAO APPROVAL

1. NAME: [redacted] b6
   (Type or print full name)

2. DATE: 27 Jul 2009   3. SIGNATURE: (b)(3) [redacted] b6

DECL ON:
DRV FROM:

OPR: MS&O/AS&O/MSS&ART

NP 11-31, DEC 08 (DS-EF)

PREVIOUS EDITIONS ARE OBSOLETE

RCS: Refer to Item 205

b6

(b)(6)

From: b6 (b)(3)
Sent: Tuesday, July 28, 2009 9:39 AM
To: FOIA (IART)
Cc: b6 (b)(3)
Subject: RE: PRIORITY -- FOIA Case #F09-0063 -APPEAL --- UNCLASSIFIED//FOUO
Attachments: F09-0063.xfdl
Classification: UNCLASSIFIED//FOUO

classification: UNCLASSIFIED//FOUO

b6 (b)(3)

No records found in DDMs relating to "magnetic resonance imaging"

Thank you,
b6 (b)(3)
b6 (b)(3)
DDMS/CoS
(b)(3)

b6 From: (b)(3)
Sent: Wednesday, July 22, 2009 1:05 PM
b6 To: (b)(3)
b6
b6 Cc: (b)(3)   SE-DAG-INBOX; ast-dag; FOIA (IART)
Subject: PRIORITY -- FOIA Case #F09-0063 -APPEAL --- UNCLASSIFIED//FOUO
Importance: High

classification: UNCLASSIFIED//FOUO

(U//FOUO) The IART has received an APPEAL to a "no records" response to a 14 May, 2009 request for
"...information on functional magnetic resonance imaging...."

(U//FOUO) The IART has accepted the appeal. We request that each tasked element search for *any records
that may exist in their holdings pertaining to magnetic resonance imaging*. The search should
include hard copy and soft copy records and documents, electronic databases, and RETIRED
RECORDS. In the event that responsive document are located, we request that each element perform an intitial
review for their own equities that should be withheld from release. We will need clean copies of responsive
documents and copies treated for sensitive equities, if applicable. Please do not task review outside your
directorate; your suggestions for additional element review(s) are welcome. OS&CI, OSC, and OGC perform a
final review of all responsive documents prior to any final release determination.

(U//FOUO) In the initial processing of this request, IMINT was tasked to search for responsive documents. In
response to the requester's appeal, IMINT is performing a new search, and the IART is requesting that DDMS,
SIGINT, SE and AS&T also search for documents. While we ask that you do not task this search outside your
directorate, the IART welcomes your suggestions for other possible sources for responsive documents within the
NRO.

(U//FOUO) IMPORTANT INFORMATION ABOUT TREATMENT OF DOCUMENTS FOR RELEASE UNDER
THE FOIA:

7/28/2009

See the Case Worksheet for the FOIA exemptions that can be invoked to withhold information. The FOIA requires that reasonably segregable releasable information within classified/sensitive documents/portions of documents be released; complete redaction of classified portions of documents is usually not an appropriate treatment. Except for the classified information exemption ((b)(1)), invocation of exemptions is discretionary; if there is no harm in release, then release is appropriate. The IART is required by law to cite FOIA exemptions to justify each individual redaction in a released document. We do not require program office reviewers to cite specific FOIA exemptions, but we do need the rationale for every instance in which you recommend that information be withheld, so the IART can apply the appropriate exemptions. If anyone involved in the review process has questions about treating documents for release, please contact me.

(U//FOUO) Please record details regarding search time/methods on the Case Worksheet provided. So that we may respond to the appeal in a timely manner and mitigate the risk of litigation, we request that you handle this request as a priority, and respond to the IART no later than 30 July 2009.

(U) If you have any questions, please contact me and reference case #F09-0063(A).

(b)(3) b6

(b)(3) b6
Senior Case Analyst
IMSC/IART
(b)(3)

UNCLASSIFIED//FOUO

UNCLASSIFIED//FOUO

7/28/2009

CASE WORKSHEET

1. CASE NUMBER: F09-0063 APPEAL	2. CASE OFFICER: (b)(3) b6	3. SUSPENSE DATE: 30 Jul 2009

4. TYPE OF REVIEW:
- [✓] FREEDOM OF INFORMATION ACT
- [ ] PRIVACY ACT
- [ ] MANDATORY DECLASSIFICATION

5. TYPE OF REQUEST:
- [ ] INITIAL
- [✓] APPEAL
- [ ] COORDINATION
- [ ] REFERRAL

6. TASK THE FOLLOWING OFFICES:
- [✓] AS&T
- [ ] BPO
- [ ] BPO/Contracts
- [ ] CIO
- [ ] COMM
- [ ] COO
- [✓] DDMS
- [ ] EEO
- [ ] GED
- [ ] IG
- [✓] IMINT
- [ ] MS&O
- [ ] MS&O/IART
- [ ] OGC
- [ ] OSC/OCC
- [ ] OS&CI/Counterintelligence
- [ ] OS&CI/Policy
- [ ] OS&CI/PSD
- [ ] OSL
- [✓] SE
- [✓] SIGINT
- [ ] SO
- [ ] OTHER: _____

7. FEE CATEGORY: [ ] COMMERCIAL [ ] NEWS MEDIA/EDUCATIONAL/SCIENTIFIC [✓] OTHER

Please review for release any document(s) found in response to a Freedom of Information Act request. Mark either your concurrence with the proposed release or non-concurrence with reasoning or justification. Reasoning or justification must be supported by one of the nine POIA exemptions. See Exemptions page.

1. RECORDS FOUND: [ ] YES [✓] NO

2.	SEARCH HOURS	REVIEW HOURS	SEARCH METHOD
IS 2-4; GS 03-08; E1-E9	_____	_____	[✓] ELECTRONIC DATABASES
GS/M 09-16; 01-06	3	_____	[✓] LIBRARIES/ARCHIVES (FILE PLAN)
GS/M 16 & Above; 07-010	_____	_____	[ ] TASKED TECHNICAL EXPERTS
			[ ] SURVEYED ALL EMPLOYEES
			[ ] OTHER: (Specify) _____

3. ACTION RECOMMENDED: [ ] GRANT IN FULL [ ] GRANT IN PART [ ] DENY [ ] GLOMAR

4. REMARKS/COMMENTS: *(If additional space is needed, attach separate page.)*
No records located regarding "magnetic resonance imaging".

1. NAME: (b)(3) b6	
(Type or print full name)	b6
2. DATE: 28 Jul 2009	3. SIGNATURE: (b)(3)

DECL ON:

DRV FROM:

OPR: MS&O/ASG/IMSC/IART

NP11-31, DEC 08 (DS-EF)

PREVIOUS EDITIONS ARE OBSOLETE

RCS: Refer to Item 205

From:

Sent: Tuesday, September 01, 2009 6:45 AM

To: ast-dag; SE-DAG-INBOX

Cc:

Subject: FW: PRIORITY -- FOIA Case #F09-0063 -APPEAL (Gilbert Roman) -- UNCLASSIFIED//FOUO

Importance: High

Attachments: F09-0063_APPEAL_Case_Worksheet.xfdl

Classification: UNCLASSIFIED//FOUO

classification: UNCLASSIFIED//FOUO

Good morning all.

(U//FOUO) Forgive me if I've missed something in the flood of emails, but I don't see **AS&T** and **SE** responses on this appeal request. Can you provide an update on the status? This is an appeal case and I need to move it forward as soon as possible to try to avoid litigation.

(U//FOUO) Any questions or concerns about this request, please contact me.

Thanks.

Senior Case Analyst
ASC/IART

From:

Sent: Wednesday, July 22, 2009 1:05 PM

To:

Cc: SE-DAG-INBOX; ast-dag; FOIA (IART)

Subject: PRIORITY -- FOIA Case #F09-0063 -APPEAL -- UNCLASSIFIED//FOUO

Importance: High

classification: UNCLASSIFIED//FOUO

(U//FOUO) The IART has received an APPEAL to a "no records" response to a 14 May, 2009 request for "...information on functional magnetic resonance imaging...."

(U//FOUO) The IART has accepted the appeal. **We request that each tasked element search for any records that may exist in their holdings** *pertaining to magnetic resonance imaging.* **The search should include hard copy and soft copy records and documents, electronic databases, and RETIRED RECORDS.** In the event that responsive document are located, we request that each element perform an initial review for their own equities that should be withheld from release. We will need clean copies of responsive documents and copies treated for sensitive equities, if applicable. Please do not task review outside your directorate; your suggestions for additional element review(s) are welcome. OS&CI, OSC, and OGC perform a final review of all responsive documents prior to any final release determination.

(U//FOUO) In the initial processing of this request, IMINT was tasked to search for responsive documents. In response to the requester's appeal, IMINT is performing a new search, and the IART is requesting that DDMS, SIGINT, SE and AS&T also search for documents. While we ask that you do not task this search outside your directorate, the IART welcomes your suggestions for other possible sources for responsive documents within the NRO.

(U//FOUO) **IMPORTANT INFORMATION ABOUT TREATMENT OF DOCUMENTS FOR RELEASE UNDER THE FOIA:**
See the Case Worksheet for the FOIA exemptions that can be invoked to withhold information. The FOIA requires that reasonably segregable releasable information within classified/sensitive documents/portions of documents be released; complete redaction of classified portions of documents is usually not an appropriate treatment. Except for the classified information exemption ((b)(1)), invocation of exemptions is discretionary; if there is no harm in release, then release is appropriate. The IART is required by law to cite FOIA exemptions to justify each individual redaction in a released document. We do not require program office reviewers to cite specific FOIA exemptions, but we do need the rationale for every instance in which you recommend that information be withheld, so the IART can apply the appropriate exemptions. If anyone involved in the review process has questions about treating documents for release, please contact me.

(U//FOUO) Please record details regarding search time/methods on the Case Worksheet provided. So that we may respond to the appeal in a timely manner and mitigate the risk of litigation, we request that you handle this request as a priority, and respond to the IART no later than 30 July 2009.

(U) If you have any questions, please contact me and reference case #F09-0063(A).

Senior Case Analyst
IPSC/IART

UNCLASSIFIED//FOUO

// b3
    b6

From: b3
           b6

Sent: Tuesday, September 01, 2009 6:51 AM

To: b3
         b6

Cc: FOIA (IART)

Subject: Gilbert Roman and F09-0063 Appeal/litigation — UNCLASSIFIED//FOUO

Classification: UNCLASSIFIED//FOUO

classification: UNCLASSIFIED//FOUO

b3 b6

(U//FOUO) Good morning, [b3 b6] This is just a brief update on where we stand on the processing of Mr. Gilbert's appeal of our "no records" response in F09-0063. I'm still waiting for a search/review response from a couple of the tasked De/Os. So far, IMINT, DDMS, and SIGINT have located no documents responsive to his request for "information on functional magnetic resonance imaging." I'm trying to corral the remaining responses so I can move a final package forward.

b3
b6

b3
b6
Senior Case Analyst
L&SC/IART
b3

UNCLASSIFIED//FOUO

2/3/

From: b3 b6

Sent: Tuesday, September 01, 2009 10:27 AM

To: b3 b6   SE-DAG-INBOX   b3 b6   ast-dag

Cc: FOIA (IART)

Subject: FW: Gilbert Roman and F09-0063 Appeal/litigation — UNCLASSIFIED//FOUO

Classification: UNCLASSIFIED//FOUO

classification: UNCLASSIFIED//FOUO

We are under the gun to produce a response for General Counsel in this FOIA Appeal. This search should be SIMPLE!!! We need your response TODAY. . . Mr. Moffett has directed me to contact the Director of Security in those offices that have not responded or he will personally call the Directorate head . . . I really don't want that to happen, so please help us out.

This gentleman is currently litigating for our failure to respond . . . push this one to the top of the list and let's get it taken care of today, please. THANK YOU!

b3 b6

Lead

FOIA/Privacy/Prepub

NSO/ASG/IMSC/IART

b3

Don't Forget to Tell Us How We're Doing...
Click Here to Open an

ASG/IMSC CUSTOMER SURVEY
We Look Forward to Hearing From You

[handwritten: Simple Search]

b3 b6

From: b3 b6

Sent: Tuesday, September 01, 2009 6:51 AM

To: b3 b6

Cc: FOIA (IART)

Subject: Gilbert Roman and F09-0053 Appeal/litigation — UNCLASSIFIED//FOUO

classification: UNCLASSIFIED//FOUO

(U/FOUO) Good morning, [b3 b6] This is just a brief update on where we stand on the processing of Mr. Gilbert's appeal of our "no records" response in F09-0063. I'm still waiting for a search/review response from a couple of the tasked Ds/Os. So far, IMINT, DDMS, and SIGINT have located no documents responsive to his request for "information on functional magnetic resonance imaging." I'm trying to corral the remaining responses so I can move a final package forward.

(b)(3)   b3

(b)(3)   b3 b6

Senior Case Analyst

IMSC/IART

(b)(3)   b3

UNCLASSIFIED//FOUO

UNCLASSIFIED//FOUO

NRO COORDINATION SHEET

NRO NUMBER: _____

#			#		
1	AS&T		13	IG	
2	BPO		14	IMINT	
3	CIO		15	MS&O	
4	COMM		16	OHR	
5	COO		17	OS&CI	APPROVE
6	COO/PC		18	OSL	
7	COO/SE		19	SIGINT	
8	DDMS		20	SO	
9	EEO/DM		21	DES	
10	GC	APPROVE	22	DDNRO	
11	GED		23	PDDNRO	
12			24	DNRO	

1. SURNAME OF ACTION OFFICER:	2. SYMBOL: IMSC/IART	3. PHONE:	4. TYPIST:	5. SUSPENSE DATE:

6. SUBJECT: FOIA Appeal - F09-0063	7. DATE:

8. SUMMARY (See Instructions Page for items that must be included in this section):

PURPOSE:
Obtain concurrence on upholding NRO's "no records" response, and obtain Appellate Authority signature on appeal response letter.

BACKGROUND: (Use continuation page, if necessary)
Please see case summary, enclosed.

RECOMMENDATION:
The IART recommends concurrence on response for Appellate Authority signature, upholding the intitial "no records" determination.

1. D&O SA SIGNATURE: _____	2. D&O ACTION NUMBER: _____
3. COS SIGNATURE: _____	4. D&O SECURITY REVIEW: _____

DECL ON:

DRV FROM:

10.2.09

OPR: EDF

PAGE   OF

PREVIOUS EDITIONS ARE OBSOLETE

NP11-21A, AUG 09 (IMT-EF)

RCS: 101-1

16 September 2009

F09-0063 APPEAL
Case Summary

- (U) Original request from requester, dtd. May 14 2009, rec'd. in the IART on 20 May **TAB 1**

- (U) NRO letter to requester, 21 May 2009, addressing specificity and fee issues **TAB 2**

- (U) Copy of court documents, dtd. 27 May 2009, sent to the IART by the requester. **TAB 3**

- (U) NRO acceptance letter, dtd. 16 June 2009 **TAB 4**

- (U//~~FOUO~~) IMINT searched for records. The NRO sent the requester a "no records" response dtd. 1 July 2009. **TAB 5**

- (U) Package from requester, rec'd in the IART 2 July 2009, re: civil action **TAB 6**

- (U) Package from United States Attorney, Eastern District of New York, re: Civil Action, dtd. 30 June 2009, rec'd. in the IART on 7 July 2009. **TAB 7**

- (U) Appeal letter from Gilbert Roman, dtd. 12 July 2009, rec'd. in the IART on 21 July, appealing NRO's "no records" response to his request for information on functional magnetic imaging **TAB 8**

- (U//~~FOUO~~) On appeal, IMINT, SIGINT, DDMS, AS&T, and SE were tasked to search for records; none of the tasked elements located responsive documents. **TAB 9**

- (U) Proposed appeal response, upholding initial "no records" determination **TAB 10**

(U) The IART requests concurrence on upholding the initial "no records" determination.

Re: F09-0063 Appeal

As an Appeal Case, this one should make its way through to Bob Harney for his approval as part of the Appeal Panel.

Thanks.

b6
(b)(3)

From: b6 (b)(3)
Sent: Tuesday, October 06, 2009 8:35 AM
To: b6 (b)(3)
Cc: FOIA (IART)
Subject: RE: Gilbert Roman and F09-0063 Appeal/litigation --- UNCLASSIFIED//FOUO
Attachments: F09-0063_Appeal Final.doc; F09-0064_APPEAL_Uphold_DIF.doc
Classification: UNCLASSIFIED//FOUO

classification: UNCLASSIFIED//FOUO
b6
Good morning, (b)(3)

(U) The Appeal response letters for F09-0063 and F09-0064 are attached. Let me know if you need anything additional..... thanks for all your help!
b6
(b)(3)

(b)(3) b6
Senior Case Analyst
IMSC/IART
(b)(3)

From: (b)(3)
Sent: Tuesday, October 06, 2009 7:55 AM
To: (b)(3) b6
Cc: FOIA (IART); Moffett Page P NRO USA CIV
Subject: RE: Gilbert Roman and F09-0063 Appeal/litigation --- UNCLASSIFIED//FOUO

classification: UNCLASSIFIED//FOUO
b6
(b)(3)

FOIA Appeal - F09-0063--GC concurs. Please send me soft copy of the letter for Mr. Barlow's signature and we will take care of from here. FYI-We are meeting w/Mr. Barlow re: Appeal F09-0064 on Tues. 13 Oct and we will take care of this one then also.

Thanks much!

b6
(b)(3)
Office of General Counsel
(b)(3)

10/6/2009

ь6

From: (b)(3)
Sent: Tuesday, September 01, 2009 6:51 AM
ь6 **To:** (b)(3)
Cc: FOIA (IART)
Subject: Gilbert Roman and F09-0063 Appeal/litigation --- UNCLASSIFIED//FOUO

classification: UNCLASSIFIED//FOUO

ь6

(U//FOUO) Good morning, (b)(3) This is just a brief update on where we stand on the processing of Mr. Gilbert's appeal of our "no records" response in F09-0063. I'm still waiting for a search/review response from a couple of the tasked Ds/Os. So far, IMINT, DDMS, and SIGINT have located no documents responsive to his request for "information on functional magnetic resonance imaging." I'm trying to corral the remaining responses so I can move a final package forward.

ь6
(b)(3)
ь6
(b)(3)
Senior Case Analyst
IMSC/IART
(b)(3)

UNCLASSIFIED//FOUO

UNCLASSIFIED//FOUO

UNCLASSIFIED//FOUO

10/6/2009

One Hundred Tenth Congress
U.S. House of Representatives
Committee on Homeland Security
Washington, DC 20515

September 26, 2007

Honorable David E. Price
Chair
Subcommittee on Homeland Security
Committee on Appropriations
Room H-218, The Capitol
Washington, D.C. 20515

Honorable Harold Rogers
Ranking Member
Subcommittee on Homeland Security
Committee on Appropriations
Room H-218, The Capitol
Washington, D.C. 20515

Dear Chairman Price and Ranking Member Rogers:

We are writing to clarify our position, as Members of the Committee on Homeland Security, with respect to funding the Office of Intelligence and Analysis in advance of the forthcoming conference for the Fiscal Year (FY) 2008 Department of Homeland Security Appropriations Act.

In an August 10, 2007 letter, Chairman Bennie G. Thompson (D-MS) expressed support for the Senate recommendation of $306 million for I&A to provide adequate resources for, among other things, the State, Local and Regional Fusion Center Initiative and compliance with the Implementing Recommendations of the 9/11 Commission Act of 2007. Since that time, the Committee on Homeland Security became aware of I&A's plans to administer the National Applications Office (NAO) on October 1, 2007. As you know, our Committee held a hearing on September 6, 2007, titled, "Turning Spy Satellites on the Homeland" that examined the privacy and civil liberties implications of the NAO at the Department of Homeland Security. The NAO will oversee the use of spy satellites for domestic purposes – including, for the first time, the provision of detailed satellite imagery to State, Local, and Tribal law enforcement. The NAO marks a dramatic expansion of prior domestic use of satellite imagery that raises very significant Constitutional, legal, and organizational issues.

While we believe the NAO may hold significant promise in helping to secure the homeland from future terrorist attacks, we are gravely concerned by the Department's lack of progress in creating the appropriate legal and operational safeguards necessary for ensuring that military spy satellites do not become the "Big Brother in the Sky" that some in the privacy and civil liberties community have described. Accordingly, the Committee on Homeland Security,

1

like the House Homeland Security Appropriations Subcommittee, have asked the Department to provide a written legal framework for the NAO and the standard operating procedures (SOPs) under which it will operate in order to allow Members an opportunity to review the plans and suggest changes to ensure that the Constitutional rights of all Americans are protected.

In the last three years, at least four programs – including the $140 million Secure Flight Program; the $100 million Computer Assisted Passenger Prescreening System II (CAPPS II) program; the $42 million Analysis, Dissemination, Visualization, Insight and Semantic Enhancement (ADVISE) program; and the $8 million Multistate Anti-Terrorism Information Exchange Pilot Project (MATRIX) – have been either canceled or suspended by the Department as a result of its failure to adhere to applicable privacy rules and regulations. We appreciate that Democrats on the House Homeland Security Appropriations Subcommittee played a critical role in bringing to light the vulnerabilities of these programs. Each of these programs, if implemented, would have compromised the privacy rights of hundreds of thousands, if not millions, of Americans. We do not want the Department to repeat the same mistakes with this program. Given the gravity of the privacy and civil liberties issues in play with the NAO, we respectfully request that the forthcoming Conference Committee adopt the House position on NAO unless and until the Department has completed its written legal framework and SOP documents and Congress has had an opportunity to review those documents and to assess their adequacy.

We look forward to working with you throughout this process to ensure that the issues identified as matters of concern by the Committee on Homeland Security and Chairman Price of the Appropriations Committee's Subcommittee on Homeland Security are resolved.

Sincerely,

Bennie G. Thompson
Chairman

Jane Harman
Chair
Subcommittee on Intelligence, Information
Sharing, and Terrorism Risk Assessment

Loretta Sanchez

Edward J. Markey

Peter A. DeFazio

Eleanor Holmes Norton

September 26, 2007
Page 3

Zoe Lofgren

Sheila Jackson-Lee

Donna M. Christensen

Bob Etheridge

James R. Langevin

Henry Cuellar

Yvette D. Clarke

Christopher P. Carney

Al Green

Ed Perlmutter

Bill Pascrell, Jr.

cc: The Honorable Dave Obey, Chairman, Committee on Appropriations
    The Honorable Jerry Lewis, Ranking Member, Committee on Appropriations

HAARP Explanation

HAARP activist concerns on implanting thoughts are supported. One name we can call it is the accelerated learning program. This program helps teach the brain to learn faster. The program states a 100 percent increase in learning. The brain is stimulated in a way that allows it to learn faster.

This is just one application we need to keep track on—HAARP. Another application of HAARP is weather manipulation. I have seen records that stated there are fifteen HAARP research facilities. I have begun the investigation on where these facilities may be. Why do we need fifteen research facilities?

If you check you tube, you will find much more information on HAARP. There are also more detailed books on this matter. Our constitution was established to protect the people from tyranny. Too many technologies are being invented and not enough public knowledge on how they are being used. We must watch and ask how these technologies are really being used. All records should be opened for public review.

The Quantum Xrroid Machine

The Quantum Xrroid (QX) machine invented by Prof. William Nelson holds many applications. If explained to me properly, it has thousand of applications. The QX machine is a biofeedback device that can record a person's electromagnetic pulses. These pulses are distinctive to the individual, just like fingerprints to each person and facial recognition systems.

We have a way to track any person in the world just by their electromagnetic signature. So we will be able to find a lost child or a person of interest anywhere in the world at any time. No more lost and missing children. With new technology, new questions will be asked. Who will collect these readings? Are they already being collected without our knowledge and approval. The people have a right to know and approve on who, what, when, where, and why this technology is to be used.

I was scanned by this device. It told me I hurt myself in a particular area of my body. This was exactly true and correct. It told me that I have been worked on before. I never knowingly had any scan done to my body before. So this means that because of all my research I became a person of interest to someone. I hope to retrieve these fills one day on how I was worked on before.

Professor Nelson an ex-NASA scientist has left the United States because our country and medical society does not allow this device to be used in the States. He has moved to Hungary to further his research. You could look all this up on the Internet.

What kind of society will we allow and leave to our children?

New Request in the Works

You shall find some of my new request in the next pages. Please read them.

National Aeronautics and Space Administration

Headquarters
Washington, DC 20546-0001

August 16, 2010

Gilbert Roman

REF: FOIA Request 10-HQ-F-01392

Dear Mr. Gilbert Roman:

Thank you for your Freedom of Information Act (FOIA) request dated July 12, 2010 and received in our office on July 20, 2010. Your request was for:

1. Active files on the HAARP program based in Alaska, on implanting thoughts and causing behavior.
2. The date of the first person it was used against successfully (HAARP) implanting thoughts and causing behavior.
3. The files on the way it has been used, to include any military operations like the invasion of Iraq, etc.
4. Files on traveling to distant planets using the gravitational pulls of these planets to propel travel at great speeds. Any working models being built or already built.
5. Files on teleportation, making objects or atoms move from one location to another.
6. Files on Quantum Xrriod-QXCI Bio-feedback machine, Professor William Nelson, Quantum biology, energetic medicine, homeopathy naturopathy.
7. Files on making objects move using some sort of electro-magnetic-FMRI technology.
8. Files on Functional Magnetic Resonance Imaging-FMRI, date of invention, date of first person used against successfully.

Since we must search for and collect documents from offices other than the office processing the request, we are availing ourselves of the 10 working day extension of response time provided in the NASA FOIA regulations (14 CFR § 1206.101 (f)(1)).

We will send a final reply to you on or before August 31, 2010.

Sincerely,

Denise Young
Headquarters FOIA Officer

7

National Aeronautics and Space Administration

Headquarters
Washington, DC 20546-0001

10-HQ-F-01392

July 21, 2010

Gilbert Roman
P.O. 170109
Ozone Park, NY 11417

Dear Mr. Roman:

This is an acknowledgement response to your Freedom of Information Act (FOIA) request, dated July 10, 2010, for "*1) active files on the HAARP program based in Alaska, 2) the date of the first person it was used against successfully (HAARP) implanting thoughts and causing behavior. 3) the files on the ways it has been used, to include any military operations like the invasion of Iraq, etc.; 4) files on traveling to distant planets; 5) files on teleportation; 6) files on Quantum Xrriod-QXCI Bio-feedback machine, Professor William Nelson, Quantum biology, energetic medicine, homeopathy naturopathy; 7) files on making objects move using some sort of electro-magnetic-FMRI technology; 8) files on Functional Magnetic Resonance Imaging FMRI.*"

Your request has been received by the National Aeronautics and Space Administration (NASA), Headquarters FOIA Requester Service Center, on July 20, 2010. We will conduct records searches with the program offices which may have responsive records.

Your request has been assigned FOIA Case Number: **10-HQ-F-01392**. All requests are processed in chronological order based upon the date it was received. NASA processes all FOIA requests in a multi-track processing system, based upon the date of receipt and the amount of work and time involved in processing the request.

FOIA processing fees may be assessed during the search and reviewing of your request; also duplication fees could also be assessed in accordance with our published agency FOIA regulations. If these fees are above the threshold of the amount you originally agreed to then we will contact you prior to any additional processing is done, to reassess your options regarding these fees.

If you would like an update on your status please contact Ms. Mary Bell, at (202) 358-1708.

Sincerely,

ORIGINAL SIGNED BY
Denise Young
Headquarters FOIA Officer

Gilbert Roman
PO box 170109
Ozone pk., NY 11417
Nov. 4, 2010

Re: Air Force Historical Research Agency

   This request is made under the freedom of information act. This request should be made free of charge but if you can not, I will authorize 100 dollars of search time per month, when cleared by me. My request is as follows:

1. Unexplained energy readings or sources in and around religious institutions. In FMRI, Quantum Xrroid, infra-red, facial recognition, full spectrum cameras, or radar.
2. Unexplained energy readings or sources in and around the rest of the earth. In FMRI, Quantum Xrroid, infra-red, facial recognition, full spectrum cameras, or radar.
3. Active files on the HAARP program, based in Alaska, on implanting thoughts and may cause behavior.
4. A present list of HAARP research facilities, how many, and where are they?
5. The date of the first person HAARP was used against successfully implanting thoughts, that may cause behavior.
6. The files on the way it has been used, HAARP, implanting thoughts that may cause behavior, to include any military operations, like the invasion of Iraq, etc...
7. Files on teleportation, making objects or atoms move from one location to another.
8. Files on traveling to distant planets using the gravitational pulls of these planets to propel travel at great speeds. Any working models being built or already built.
9. Files on Quantum Xrroid machine-bio feed back, invented by Prof. William Nelson, quantum biology, energetic medicine, homeopathy.
10. Files on making objects move using some sort of electric-magnetic-FMRI technology.
11. Files on Functional Magnetic Resonance Imaging-FMRI, date invented, date of first person used against successfully.

Gilbert Roman

DEPARTMENT OF THE AIR FORCE
HEADQUARTERS, UNITED STATES AIR FORCE
WASHINGTON, DC

2 September 2010

HAF/IMIO (FOIA)
1000 Air Force Pentagon
Washington DC 20330-0100

Mr. Roman Gilbert

Ozone Park NY 11417

Dear Mr. Gilbert

This is in response to your Freedom of Information Act (FOIA) request, dated July 10, 2010 for information relating to HAARP.

The Air Force Historical Research Agency (AFHRA), AFHRA/RSA, 600 Chennault Circle, Maxwell AFB AL 36112-6424 provided the following list of documents below that maybe responsive to your request. AFHRA request you review the listing and identify which items you desire and contact them directly at the address listed above. They will provide you with a response and a FOIA request is not required with their agency.

AFHRA provided the following:

Main: 0011 AIR FORCE
RECTYPE: Periodic History   Call: K480.01 V.7
IRIS Number: 1104467
BegDate: 01-01-1991   EndDate: 12-31-1991   Author: ALASKAN NORAD REGION
Title Extension: VOL VII OF XII
Title Added Entries: INTERVIEW, HISTORIAN WITH COL NEIL R. MCCOY, PLANS, 11 AIR FORCE (AF), 1 APR 92 OVER THE HORIZON BACKSCATTER (OTH/B) (RADAR) CAPABILITIES DEMONSTRATION REPORT, 18 JUN 91 TERMINATION PLAN, ALASKA RADAR SYSTEM (ARS) OVER THE HORIZON BACKSCATTER RADAR PROJECT, JUL 91 HAARP, HIGH FREQUENCY ACTIVE AURORAL RESEARCH PROGRAM, GEOPHYSICS LABORATORY, OFFICE OF NAVAL RESEARCH, FEB 90 ENVIRONMENTAL ASSESSMENT: PROPOSED BEDDOWN OF F-15E SQUADRON, ELMENDORF AIR FORCE BASE (AFB) AK, FEB 91
Class: SECRET
Abstract: AFTER ACTION REPORTS, 3115 OPERATIONS PLAN (OPLAN). LETTER OF AGREEMENT BETWEEN HEADQUARTERS, 11 AIR FORCE AND OVER THE HORIZON RADAR SYSTEMS PROGRAM OFFICE, ELECTRONIC SYSTEMS DIVISION, OUTLINING MAJOR RESPONSIBILITIES OF 11 AF, PACIFIC AIR FORCES (PACAF), AND OVER THE HORIZON RADAR SYSTEMS PROGRAM OFFICE CONCERNING RESIDUAL ACTIONS REQUIRED BY CANCELLATION OF

PROGRAM. INFORMATION ON ALASKAN UNATTENDED RADAR STATION (UAR)
FACILITIES.
Admin: Warning Notice, Intelligence Sources or Methods Involved Originating Authority
Determination Requested -- OR -- Other administrative markings
DNotes: SUPPORTING DOCUMENTS 3-10 THROUGH 3-82.
Reel: 44291   Frame: 1189   FrameLast: 1674

Main: 0011 AIR FORCE
RECTYPE: Periodic History   Call: K480.01 V.1
IRIS Number: 1104461
BegDate: 01-01-1991   EndDate: 12-31-1991   Author: ALASKAN NORAD REGION
Title Extension: VOL I OF XII
Class: SECRET
Abstract: LT GEN THOMAS G. MCINERNEY SERVED AS COMMANDER, 11 AIR
FORCE (AF). 11 AF OPERATED THREE AIR FORCE BASES: EIELSON,
ELMENDORF, AND SHEMYA; TWO FORWARD OPERATING BASES (FOB)
(GALENA AND KING SALMON); AND TWENTY RADAR SITES. UNITS INCLUDED 3
WING (WG), 11 TACTICAL CONTROL WING (TCW), 343 WG, 5073 AIR BASE
GROUP (ABG), AND 623 AND 643 SUPPORT SQUADRONS (SS). MCINERNEY
SERVED AS COMMANDER, ALASKAN COMMAND (ALCOM), SUBORDINATE
UNIFIED COMMAND REPORTING TO UNITED STATES PACIFIC COMMAND
(USPACOM); COMMANDER, ALASKAN NORAD REGION; AND COMMANDER,
JOINT TASK FORCE/ALASKA (JTF/AK), JOINT NATURAL CONTINGENCY
COMMAND. 21 WG INACTIVATED; REPLACED BY 3 WG REASSIGNED FROM
CLARK AIR BASE (AB), PHILIPPINES. 11 TACTICAL AIR SUPPORT SQUADRON
(TASS) AND 11 TACTICAL INTELLIGENCE SQUADRON (TIS) ACTIVATED;
ELMENDORF REGIONAL HOSPITAL REDESIGNATED 11 MEDICAL CENTER.
MEDICAL CENTER ASSUMED RESPONSIBILITY FOR MEDICAL SUPPORT TO
PACIFIC AREA PREVIOUSLY PROVIDED FROM CLARK AB. 752 AIR FORCE BAND
REDESIGNATED AIR FORCE BAND OF THE PACIFIC. 1931 COMMUNICATIONS
GROUP (CG) AND 71 AIR RESCUE SQUADRON (ARS) INACTIVATED. 210 ARS OF
ALASKA AIR NATIONAL GUARD ASSUMED SEARCH AND RESCUE MISSION.
PREPARATIONS MADE FOR INACTIVATION OF 6 STRATEGIC RECONNAISSANCE
WING (SRW). PACIFIC AIR FORCES (PACAF) CONDUCTED UNIT EFFECTIVENESS
INSPECTION (UEI) OF 11 TCW, 5073 ABG, ALL FOB AND OTHER SUPPORT
ORGANIZATIONS 21 MAR THROUGH 17 APR 91. 17 TAS AIRLIFTED CHRISTMAS
PRESENTS TO ISOLATED NATIVE COMMUNITY, ARCTIC VILLAGE. AERIAL
DEMONSTRATION BY NAVY BLUE ANGELS HIGHLIGHTED ARMED FORCES DAY
OPEN HOUSE AT ELMENDORF. 11 AF INSTALLATIONS HOSTED VICE
PRESIDENT DAN QUAYLE; SEVERAL UNITED STATES SENATORS AND
REPRESENTATIVE PAT SCHROEDER; SECRETARY OF DEFENSE RICHARD
CHENEY; SOVIET VISITORS, HIGHEST RANKING WAS GEN COL ALEKSANDR V.
KOVTUNOV, COMMANDER IN CHIEF OF SOVIET FAR EAST MILITARY THEATER
OF OPERATIONS. ALASKA BASED AIR FORCE PERSONNEL SUPPORTED
OPERATION DESERT STORM BY DEPLOYING DIRECTLY INTO THEATER OR
SUPPLEMENTING FORCES IN PACAF OR CONTINENTAL UNITED STATES
(CONUS). 11 AF EXPERIENCED LARGE BUDGET CUTS. CHERRY HILL MILITARY

FAMILY HOUSING AREA ON ELMENDORF COMPLETED. ALASKA RADAR SYSTEM (ARS) OR ALASKA OVER THE HORIZON BACKSCATTER (OTH/B) RADAR TERMINATED. HIGH FREQUENCY (HF) ACTIVE AURORAL RESEARCH PROGRAM (HAARP) CONSIDERED AS ALTERNATIVE USE FOR POWER PLANT BUILT NEAR GULKANA. SECOND PHASE OF NORTH WARNING SYSTEM (NWS) IMPLEMENTED. NEGOTIATIONS CONTINUED ON TURNOVER TO FEDERAL AVAIATION ADMINISTRATION (FAA) OF LONG RANGE RADARS (LRR) AT FORT YUKON AND INDIAN MOUNTAIN. BEDDOWN OF F-15E AIRCRAFT AT ELMENDORF AND F-16 AIRCRAFT AT EIELSON BEGAN. UNITS PARTICIPATED IN NOISE AND SONIC BOOM IMPACT TECHNOLOGY STUDY. 11 AF ASSISTED IN COUNTER NARCOTICS MISSIONS. EXERCISES INCLUDED ARCTIC WARRIOR, FREQUENT OBSERVER, AND COPE THUNDER NORTH.
Admin: Not Releasable to Foreign Nationals Warning Notice, Intelligence Sources or Methods Involved Originating Authority Determination Requested -- OR -- Other administrative markings
Reel: 44289   Frame: 1429   FrameLast: 1862

Main: 0011 AIR FORCE
RECTYPE: Periodic History   Call: K480.01 V. 12
IRIS Number: 1115818
BegDate: 01-01-1993   EndDate: 12-31-1993   Author: ALASKAN NORAD REGION ** WING/0003
Title Extension: VOL XII OF XII
Title Added Entries: DEPARTMENT OF AIR FORCE, RECORD OF DECISION, HIGH FREQUENCY ACTIVE AURORAL RESEARCH PROGRAM (HAARP), FINAL ENVIRONMENTAL IMPACT STATEMENT, GAKONA, ALASKA, 18 OCT 1993 ** FEDERAL AVIATION ADMINISTRATION (FAA), PRELIMINARY DRAFT REPORT, JOINT SURVEILLANCE SYSTEM ALASKA, MINIMALLY ATTENDED RADAR STUDY, 31 DEC 1993 ** 11 AIR FORCE (11 AF), LETTER, SEARCH AND RESCUE (SAR) SUMMARIES, 1993 ** 343 OPERATIONS GROUP (OG), MESSAGE, COPE THUNDER 93-1, 93-2, 93-3, 93-4, MAY-SEP 1993 ** 343 WING, EXECUTIVE SUMMARY, AIRCRAFT ACCIDENT INVESTIGATION CONDUCTED IN ACCORDANCE WITH AFR 110-14, F-16D S/N SN NO. 89-2069, EIELSON AIR FORCE BASE. AK, 18 FEB 1994
Class: SECRET
Admin: For Official Use Only
DNotes: SUPPORTING DOCUMENTS III-30 THROUGH III-89. FULL TEXT DOCUMENT IS AVAILABLE IN ELECTRONIC FORM VIA CLASSIFIED NETWORK.
Reel: 45964   Frame: 867   FrameLast: 1477

Main: 0011 AIR FORCE
RECTYPE: Periodic History   Call: K480.01 V. 10
IRIS Number: 1115816
BegDate: 01-01-1993   EndDate: 12-31-1993   Author: ALASKAN NORAD REGION ** WING/0003
Title Extension: VOL X OF XII

Title Added Entries: ALASKAN NORAD REGION (ANR)/ALASKAN AIR COMMAND (AAC), REGULATION 60-1, POSITIVE CONTROL OF AIRCRAFT OPERATING IN AIRSPACE ADJACENT TO SOVIET UNION, 9 SEP 1993 ** 11 AIR FORCE (11 AF), LETTER, REAL ESTATE ISSUES IN ALASKA, 11 MAY 1993 ** PHILLIPS LAB, BRIEFING, HIGH FREQUENCY ACTIVE AURORAL RESEARCH PROGRAM (HAARP), 22 JUL 1993 ** PHILLIPS LAB, LETTER, SEMI-MONTHLY ENVIRONMENTAL IMPACT ANALYSIS PROCESS STATUS LETTER, JAN-NOV 1993
Class: SECRET
Admin: For Official Use Only;Not Releasable to Foreign Nationals;Warning Notice, Intelligence Sources or Methods Involved
DNotes: SUPPORTING DOCUMENTS III-1 THROUGH III-28. FULL TEXT DOCUMENT IS AVAILABLE IN ELECTRONIC FORM VIA CLASSIFIED NETWORK.
Reel: 45963   Frame: 718   FrameLast: 1460

Main: 0011 AIR FORCE
RECTYPE: Periodic History   Call: K480.01 V. 7
IRIS Number: 1115802
BegDate: 01-01-1992   EndDate: 12-31-1992   PubDate: 12-08-1993   Author: ALASKAN NORAD REGION
Title Extension: VOL VII OF X (XI)
Title Added Entries: 11 AIR FORCE (11 AF),STAFF SUMMARY SHEET, OTH-B (OVER-THE-HORIZON BACKSCATTER) POWER PLANT AND HAARP (HIGH FREQUENCY ACTIVE AURORAL RESEARCH PROGRAM) NEGOTIATIONS, 6 FEB 1992 ** PHILLIPS LAB AND OFFICE OF NAVAL RESEARCH, HAARP FACT SHEET, 10 AUG 1992 ** PHILLIPS LAB, LETTER, HAARP BI-MONTHLY ENVIRONMENTAL IMPACT ANALYSIS PROCESS, JUN-DEC 1992 ** 11 AF, BRIEFING, COMMAND POST STUDY, ELMENDORF AIR FORCE BASE (AFB), AK, AS OF 31 DEC 1992 ** 11 AF, AFTER ACTION REPORT, RUSSIAN AN-124, 23 MAR 1992 ** 11 AF, AFTER ACTION REPORT, RUSSIAN VISIT, 25 APR 1992 ** 11 AF, ITINERARY, RUSSIAN SU-27 AND AN-124, 12 JUL 1992 ** 11 AF, SUMMARIES, SEARCH AND RESCUE SUMMARIES, JAN-DEC 1991 ** 11 AF, TALKING PAPER, COMMONWEALTH OF INDEPENDENTS STATES (CIS)/UNITED STATES (US) SAREX (SEARCH AND RESCUE EXERCISE), N. D. ** JOINT REPORT OF US COAST GUARD/UNITED STATES AIR FORCE (USAF)/CIS SAREX AND EXCHANGE OF SUBJECT MATTER EXPERTS, N. D. ** 11 AF, REPORT, GROUND SAFETY MISHAP ANALYSIS, A FIVE YEAR COMPARISON, 1 JUL 1992
Class: SECRET
Admin: For Official Use Only;Not Releasable to Foreign Nationals;Warning Notice, Intelligence Sources or Methods Involved
DNotes: SUPPORTING DOCUMENTS III-17 THROUGH III-24 AND IV-1 THROUGH IV-45. FULL TEXT DOCUMENT IS AVAILABLE IN ELECTRONIC FORM VIA CLASSIFIED NETWORK.
Reel: 45959   Frame: 771   FrameLast: 1267

Main: ALASKAN COMMAND
RECTYPE: Periodic History   Call: K484.011 V. 1
IRIS Number: 1121972

BegDate: 01-01-1996   EndDate: 12-31-1996   PubDate: Author: CLOE, JOHN H.
** FRANK, JAMES R., JR. ** GOHL, STAN
Title Extension: VOL I OF III
Class: SECRET
Subject: ELMENDORF AFB, AK
Abstract: ALASKAN COMMAND (ALCOM) DEVELOPED JOINT MISSION ESSENTIAL
TASK LIST (JMETL). PLANNED FOR CLOSURE OF FORT GREELY AND ADAK
NAVAL AIR FACILITY (NAF), AK AND FOR DEVELOPMENT OF THE AREAS. LT
GEN PATRICK K. GAMBLE ASSUMED COMMAND OF ALCOM FROM LT GEN
LAWRENCE E. BOESE ON 21 AUG 1996. CONDUCTED JOINT READINESS
MOBILITY EXERCISE. REEVALUATED ANTITERRORISM MEASURES AFTER
KHOBAR TOWERS BOMBING IN SAUDI ARABIA ON 25 JUN 1996. RECEIVED
COMMUNICATIONS AND COMPUTER EQUIPMENT TO SUPPORT JOINT WORLD-
WIDE INTELLIGENCE COMMUNICATIONS SYSTEM (JWICS). HUMAN
INTELLIGENCE (HUMIT) SUPPORT PROVIDED FOR SEARCH AND RESCUE (SAR).
DISCUSSED IMAGERY SUPPORT FOR MILLER'S REACH FIRE. INCREASED
EMPHASIS ON BIRD AIRCRAFT STRIKE HAZARD (BASH) PROGRAM AFTER 22
SEP 1996 E-3B AIRCRAFT CRASH AT ELMENDORF AIR FORCE BASE (AFB), AK.
18 FIGHTER WING (FW) F- 15C AIRCRAFT CRASHED NEAR EIELSON AFB, AK
WHILE DEPLOYED TO ALCOM. PARTICIPATED IN COPE THUNDER 94-4.
SUPPORTED ALASKA DISASTER PREPAREDNESS PLAN. ALCOM IMPLEMENTED
EXERCISE POLAR BRIDGE. PARTICIPATED IN EXERCISE NORTHERN EDGE. F-
15 AIRCRAFT AND PERSONNEL VISITED RUSSIA AS GOOD WILL GESTURE.
CONDUCTED ARCTIC SAREX 96. DEFENSE FUELS AGENCY TURNED OVER
WHITTIER PIPELINE TO THE UNITED STATES ARMY. STUDIED MILITARY
LANDFILL REQUIREMENTS. WORLD WIDE MILITARY COMMAND AND CONTROL
SYSTEM (WWMCCS) TERMINATED; GLOBAL COMMAND AND CONTROL SYSTEM
(GCCS) IMPLEMENTED. PENTAGON SPONSORED HIGH-FREQUENCY ACTIVE
AURORAL RESEARCH PROGRAM (HAARP). SUPPORTED NORTHERN EDGE 96
COMMAND POST EXERCISE.
Admin: For Official Use Only Originating Authority Determination Requested -- OR --
Other Administrative Markings
DNotes: APPENDICES INCLUDE: BIOGRAPHIES OF KEY PERSONNEL; ROSTER
OF KEY PERSONNEL; ROSTER OF ALASKAN COMMAND (ALCOM) CIVILIAN
ADVISORY BOARD (CAB) MEMBERS; MILITARY CONSTRUCTION LISTING;
ALCOM PLANS LISTING.
Reel: 48123   Frame: 6   FrameLast: 143

Please contact the undersigned at (703) 693-2736 should you have any questions
and refer to case #2010-07136-F.

Sincerely

ESPINAL.JOHN.M.11   Digitally signed by ESPINAL.JOHN.M.1184810375
                      DN: c=US, o=U.S. Government, ou=DoD, ou=PKI,
84810375   ou=USAF, cn=ESPINAL.JOHN.M.1184810375
                      Date: 2010.09.02 09:29:19 -04'00'

JOHN M. ESPINAL
HAF FOIA Disclosure Officer

Evidence of the Existence of God

Is it possible that our government has proof of existence of God? This contact has been made. The god from the Holy Bible, Holy Quran, and the Torah if true, how many persons can be saved from God's final judgment?

I have begun a new investigation into this fact. Time and evidence will prove the truth. I really believe this to be true. One reason that leads me to FMRI technology is one and the same, so I have reliable points of entry into this investigation.

From a biblical aspect, it makes sense to me. In the book of Genesis, Adam and Eve eat an apple that allows them to know good and evil. That God said because you have eaten the apple you will truly die.

We have technologies that can help prove the existence of God. We have FMRI technology that can see if a person is good, evil, or a liar. Does Adam and Eve knowing good and evil have anything to do with FMRI technology? As children of God, we should have God-like abilities. I hope to prove and support the existence of God one day soon.

A priest once said, "For those that believe no proof is necessary, for those who do not, no proof is possible." I go a little further by saying, "God can prove himself at any time in history, to any person, believer, or not."

One Hundred and Eighty-three New Documents

One hundred and eighty-three documents were withheld until a couple of days after my book was released. These documents were sent out in December 2009, and I did not receive them until May 2010. In these documents, we will find further proof of the technology that reads our thoughts. In these documents, we find the NRO's claim not to have records on FMRI technology is false. We find that HAARP activist worries about implanting thoughts are very true. I have entered some of the one hundred and eighty-three documents because many pages have technical data. Please review these documents in the next pages.

OFFICE OF THE SECRETARY OF DEFENSE
1950 DEFENSE PENTAGON
WASHINGTON, DC 20301-1950

1 5 MAR 2010

ADMINISTRATION AND
MANAGEMENT

Ref: 10-L-0385

Mr. Gilbert Roman

Ozone Park, NY 11417

Dear Mr. Roman,

   This is in response to your October 22, 2009, Freedom of Information Act (FOIA) request to the Defense Advanced Research Projects Agency (DARPA) for information pertaining to Functional Magnetic Resonance Imaging (FMRI) for which you have now filed litigation. That Agency forwarded your request to the Department of Defense Office of Freedom of Information (OFOI), which handles all FOIA requests for DARPA. OFOI sent an interim response to you on December 30, 2009. On December 23, 2009, you filed a complaint in the Eastern District of New York, which was served on or about January 8th, 2010. This office was informed of the complaint on January 15, 2010.

   This office has been advised by DARPA that they have conducted a two hour search of their records, and found publicly available information concerning FMRI technology (Enclosed). Additionally, DARPA has advised this office that parts 2-4 of your request are not under their purview, and therefore have no records pertaining to those subjects. Since this request is in litigation all appeal rights are moot.

Sincerely,

Will Kammer

Will Kammer
Chief, Freedom of Information Division

Enclosure
As stated

The ROC curve is a graph of the fraction of true positives versus the fraction of false positives produced by a binary classifier, as a function of discrimination threshold. An AUC of 1.0 indicates perfect classification, with all trials from each of the two classes correctly identified, whereas an AUC of 0.0 indicates perfect misclassification. An AUC of 0.5 corresponds to a random classifier, which is just as likely to produce true as false positives. Therefore, binary classifiers with AUC statistics between 0.5 and 1.0 will result in class discrimination performance better than chance (Mason and Graham, 2002).

The false positive and false negative rate calculations generate statistics that measure discriminator sensitivity and specificity by computing the fraction of trials classified incorrectly with respect to class type. The false positive rate statistic indicates the fraction of class 0 trials (negative class) classified incorrectly as class 1 (positive class), whereas the false negative rate identifies the fraction of class 1 trials incorrectly classified as class 0.

Training and Testing Stages

During the test stage, a discriminator is trained on a sequence of attribute vectors derived from trials of known class types (i.e., those to be discriminated during the test stage). The attribute vectors, representing points in trial attribute space, should ideally cluster into two disjoint groups, each corresponding to the trial class from which they were derived. Therefore, it is important that the trial attribute vectors encode to the maximum extent possible only trial information unique to their respective classes and not information or attributes common to both classes.

The test stage estimates class membership of a new set of trial data by employing the discriminator obtained from the training stage. To maximize the discriminator performance, the procedures and settings for data acquisition, processing, and transformation to attribute space should be identical to those employed during discriminator training, at least to the extent possible.

Application to Detecting Targets in Satellite Imagery

To apply this single-trial analysis to classify individual perceptual events, we used a rapid serial visual presentation (RSVP) method with satellite images with three target types: helipads, surface to air missile (SAM) sites, and antiaircraft artillery (AAA) sites. The goal was to determine, in near real time, the brain responses that indicate when a subject has detected a target.

Information extracted from briefly viewed stimuli has long been studied using RSVP paradigms because they mimic the rapid exposure of visual stimuli to the eyes as a person scans either a picture or a sentence. In an RSVP paradigm, stimuli are presented continuously at rapid rates to subjects. Behavioral research has shown that conceptual information for each briefly presented stimulus is extracted between 100-300 ms after stimulus onset (Intraub, 1999; Potter, 1999). This rapid extraction of information is a function of what Potter refers to as conceptual short-term memory (CSTM), which is different from working memory. The main difference is that unlike working memory that engages controlled or executive cognitive processes over seconds or minutes, CSTM is capable of rapid conceptual processing.

Using the RSVP paradigm with EEG measures, Thorpe et al. (1996) demonstrated that information can be extracted from briefly presented (20 ms) pictures. Thorpe et al. asked

News Release

Media Contact:

HONEYWELL TECHNOLOGY TO HELP U.S. MILITARY RAPIDLY ANALYZE INTELLIGENCE AND KEEP TROOPS OUT OF HARM'S WAY

Brain Monitoring Sensor Technology Increases Speed and Precision to Help Analysts Identify Threats

PHOENIX, Nov. XX, 2007—Honeywell (NYSE: HON) announced today that it is developing a revolutionary system for the Defense Advanced Research Agency (DARPA) that could dramatically improve the military's intelligence analyzing capabilities by allowing analysts to evaluate images from satellites, ground cameras, and surveillance aircraft more precisely and quickly than ever before.

The Honeywell Image Triage System (HITS) will enable Department of Defense (DoD) personnel to analyze intelligence images six times faster than the current computer-based system through the use of high-tech sensors that monitor signals in the human brain. Honeywell is developing the system as part of DARPA's Neurotechnology for Intelligence Analysts (NIA) program.

"Computer-based systems currently in use cannot process enormous volumes of intelligence imagery fast enough to meet the needs the military," said Bob Smith, Vice President, Advanced Technology, Honeywell Aerospace. "That's why we are developing technology that speeds up the intelligence analysis process by tapping into brain signals associated with split-second visual judgments. As a result, we are going to give analysts the ability to identify dangerous threats to our troops more quickly, precisely and effectively than ever before."

The human brain is capable of responding to visually salient objects significantly faster than an individual's visual-motor, transformation-based response. Simply put, when examining an image an analyst's brain can register a discovery long before the analyst becomes fully aware of it.

Honeywell's technology uses sensors to monitor brain activity in real time, automatically identifying and recording brain signals to tag intelligence images worthy of additional review. The system presents data to analysts in high speed bursts of 10 to 20 images per second. Head-mounted electroencephalogram (EEG) sensors detect neural signals associated with target recognition as the images are viewed. Neural signals known as "event related potentials" are used to tag the images that contain likely targets or threats. At the end of the high-speed scan, the analysts are able to focus on the small subset of key images tagged by the brain scan instead of searching slowly and systematically through every inch of high resolution satellite images like current techniques require.

Honeywell's triage analysis methods will ultimately apply to a diverse range of imagery, including high resolution electro-optical, infrared, and video imagery. It could eventually be used in a broad range of military and commercial applications, including medical diagnosis and geospatial analysis.

"HITS is going to help the military to analyze more intelligence imagery everyday. By more quickly identifying threats to our troops, Honeywell is helping the U.S. military keep them out of harm's way," Smith said.

Honeywell International is a $34 billion diversified technology and manufacturing leader, serving customers worldwide with aerospace products and services; control technologies for buildings, homes, and industry; automotive products; turbochargers; and specialty materials. Based in Morris Township, NJ, Honeywell's shares are traded on the New York, London, and Chicago Stock Exchanges. It is one of the 30 stocks that make up the Dow Jones Industrial Average and is also a component of the Standard & Poor's 500 Index. For additional information, please visit www.honeywell.com.

Based in Phoenix, Honeywell's aerospace business is a leading global provider of integrated avionics, engines, systems, and service solutions for aircraft manufacturers, airlines, business and general aviation, military, space, and airport operations.

#   #   #

Introduction

The brain is the machinery of the mind, and an important goal for brain scientists is to understand how component mental processes (such as attention, pattern recognition, and memory) arise out of neuronal processes. Beyond the basic science questions, it is also important to translate this understanding to help diagnose disease states, and to solve practical problems by leveraging the power of modern technological methods of assessing this biological machinery. For practical applications, it has long been the dream of many researchers to use brain outputs to interface with computers as a way to help patients (such as those who have restricted mobility) to improve the quality of their lives. With programs such as augmented cognition (e.g., Schmorrow et al., 2006), it is now possible to imagine that we could improve upon the way human operators interact with complex information and control systems through direct assessment of neural activity and through feeding the interpretation of this activity back to guide the operator's workload. Researching practical methods of advancing neurotechnology has now become a priority because of the high bandwidth of information presented to human operators in many industrial or military environments. For example, continual improvement in the coverage and resolution of the U.S. Geospatial Intelligence Agency's satellite intelligence imagery presents the image analyst with increasingly rich information that is potentially critical for national security. However, this information will contribute to security only if the person can detect potentially significant objects and events within the very dense stream of visual information.

The noninvasive measurement of brain processes can be accomplished with various neurotechnologies, such as EEG, magnetoecephalography, functional MRI, and near-infrared spectroscopy, with each technology measuring different aspects of neuronal function. These technologies also differ with regard to their spatial and temporal resolution. For practical applications, it is particularly important to assess brain processes on the millisecond timescale at which cognitive function emerges from neural activity. For application in real world environments, unobtrusive and portable measures are useful. Of the neurotechnologies now available, only EEG is both portable and capable of measuring neuronal function with adequate spatial and temporal resolution. Whereas near-infrared spectroscopy (NIRS), a technology in its infancy, is in principle inexpensive and portable, and some laboratories have shown fast optical responses with the necessary temporal resolution, the signal-to-noise level requires extensive signal averaging for the fast response, making it unworkable for single trial measures with millisecond temporal resolution.

Signal averaging is required for the fast optical response, and it is typically applied to evoked or event-related potential (ERP) measures of the EEG because the signals of interest are small and must be extracted from the background noise. Limiting the measure to the average of many experimental trials is inadequate for both performance enhancement and learning contexts in which the response of the brain must be measured in relation to individual instances.

In this paper, we discuss a single-trial analytic framework for EEG measurement, and we present an example of how this method can be applied to understanding—and potentially improving—human performance in the demanding information processing task of detecting targets of military interest in a rapid stream of satellite images. In our experiment, participants were expert imagery analysts. The images (chips or segments of high-resolution broad area satellite images) were presented at ten per second. Even though the targets were centered in each chip, this rate was generally too fast for accurate conscious detection of targets. The applied example we use illustrates how outcomes of the brain's object recognition process can be analyzed and tracked trial by trial in a structured manner in a demanding military intelligence context.

Single-Trial Classification Framework

The specific strategy for this single-trial analytic framework was first to identify the neural sources that contribute to the recognition of targets (based on signal averaged responses) and then to compute measures of activity in neural source regions on a single-trial basis to use in the target classification statistics. This strategy is directed by our interest not only to detect whether the brain identifies a target, but also to understand the underlying neural mechanisms of the detection. Because artifacts such as eye movements and blinks introduce confounds into the neural source analysis process, we applied an artifact removal procedure prior to identifying neural sources, which we call Directed Components Analysis (DCA). Based on the work of Ille et al. (2002), this method not only characterizes an artifact component based on the topography of the component (e.g., eyeblink), but also conducts an ongoing components analysis of the artifact free EEG to insure that artifact correction does not subtract valid EEG activity that overlaps with the artifact template. The analytic process includes the following: signal averaging, artifact correction for noise mitigation, source estimation for the average neural responses, computation of source activity in the unaveraged data, discriminant analysis of target recognition on each trial, and, finally, performance statistics for the effectiveness of the single-trial discrimination.

Signal Averaging

The single-trial analysis method begins with signal averaging to identify relevant EEG features that distinguish the conditions of interest (e.g., target present or target absent). Signal averaging is based on the assumption that signals occur at fixed latencies after stimulus onset (i.e., phase-locked) and that noise is randomly distributed. To derive the event-related potential (ERP), conventional procedures are used, including, filtering, removing blink artifacts from the ongoing EEG, and excluding artifact contaminated trials prior to averaging. Once the average is obtained, ERP components are examined to identify those that differentiate between two and more psychological conditions (or brain states), primarily through computation of a difference wave (which can be tested for significant discrimination with t-statistics). After transformation from scalp to source space, the source activity is used to identify and select brain regions of interest, as well as to identify the time window (relative to stimulus onset) in which the brain regions are active. The result of these selections is the variables used in the single-trial target classification analysis.

Noise Mitigation

Noise mitigation, including artifact cleaning, is an important step for both signal averaging and single-trial analysis. Approaches to noise mitigation vary depending upon the noise source. One approach involves component decomposition methods, such as independent components analysis (ICA: Makeig et al., 1997). These methods can be used to remove noise introduced by various sources (e.g., ocular, muscular, and electromagnetic). Less general approaches target the nature of the noise. For example, through the use of noise reference sensors and adaptive noise cancellation methods electromagnetic interference from environmental sources can be removed effectively (Volegov et al., 2004). Specific approaches remove both noise and signals contained within particular frequencies. An example is digital filtering for removal of line-noise (50 Hz or 60 Hz).

Our single-trial analytic method relies on digital filtering and the DCA method of artifact removal to carefully minimize the distortion of brain activity (Poolman et al., 2006). The DCA method overcomes the spatial orthogonality restrictions that arise when applying traditional principal component analysis (PCA) to mixtures of artifact and brain activity topographies (see Ille et al., 2002, for an overview). In short, DCA employs the ability to continuously monitor and evaluate the temporal and spatial evolution of artifact activity in order to mark artifact-contaminated and artifact-free intervals. If an artifact topography is present in the EEG only during specific time intervals, such as eye blinks, then data acquired outside these intervals can model brain activity without fully capturing the artifact activity that may be spatially correlated with brain activity. Conversely, frames of EEG from select time points in the artifact-contaminated intervals enable the construction of templates that describe artifact topographies, but will not account for all the brain variance. Time slices are extracted from these intervals to construct artifact topographies and to form a basis for modeling artifact-free brain activity. From these, an appropriate filter can be derived to extract the artifact activity exclusively.

For the hybrid method, brain activity is modeled by a subset of orthogonal eigenvectors, derived from the covariance matrix of artifact-free EEG frames that meet two criteria: (1) they are not highly correlated with the blink template and (2) they describe substantial EEG variance. Unlike Ille et al. (1997), we set a high threshold for rejecting eigenvectors that are correlated with the blink template: only those correlated 0.95 or higher are rejected (vs. ~ 0.40 or higher for Ille et al.). Although eigenvectors nearly identical with the blink template must be rejected, those that are not should be retained. Since the eigenvectors are orthogonal, rejecting an eigenvector with moderate correlation to the blink template will result in a set of remaining eigenvectors essentially orthogonal to the blink topography and unable to accurately capture activity recorded over frontal regions (where blink activity is maximal). Brain activity not described by this reduced eigenvector set in the forward model may then load onto the blink template and be extracted by the spatial filter thereby distorting the remaining EEG in the cleaned output. Finally, the remaining eigenvectors with the largest eigenvalues that account for a specified fraction of the data's variance are retained.

The blink (or artifact) template and retained brain activity eigenvectors (cortical eigenvectors) are arranged in the columns of a matrix, as illustrated in Figure 1. The blink template's spatial filter (K^*_{EB}) is then computed as the first row vector of this matrix's pseudo-inverse. A corresponding cleaning matrix is produced by subtracting the outer product of the blink template and spatial filter from an identity matrix of equal dimension. Subsequent multiplication of the cleaning matrix with the artifact-contaminated EEG (channels × time samples) generates the blink—or artifact-free EEG.

For the cleaning matrix to maximally extract artifacts while minimizing the extraction of correlated brain activity, the blink spatial filter (K^*_{EB}) must be orthogonal to the retained cortical eigenvectors. This can only occur when the spatial filter's corresponding blink template is not in the column space of the cortical eigenvectors or, equivalently, the fraction of the normalized template's variance captured by the cortical eigenvectors is less than one. By orthogonality, the fraction of the captured normalized template variance is given by:

$$Fraction\ of\ template\ variance = \sqrt{\sum_1 \langle K_{EB} \cdot K_{BAI} \rangle^2}$$

the square root of the sum of squares of the coefficients of projection of the normalized blink template onto the cortical eigenvectors. Should the fraction equal one, the blink template will be completely described by the eigenvectors modeling brain activity, and the resulting spatial filter will extract blink-template-correlated brain activity along with artifacts.

Data Transformation—Scalp to Source

The unique step of our approach to single-trial analysis is the use of waveforms of source (i.e., cortical) activity (estimated as current density) derived from the scalp potential data. The EEG, as recorded at the scalp, represents the combined effect of activity at multiple sources. This superposition of source activities, due to volume conduction, occurs at every sensor and is a major problem for the assessment of brain function. Although different brain sources may be involved in separate cognitive processes, their activities become completely confounded at the scalp.

To avoid the superposition problem and to provide a component identification that reflects underlying neurophysiological mechanisms, the scalp-recorded data must be transformed into source space. Such a transformation attempts to separate the overlapping signals and attributes the scalp-recorded data to the relevant brain regions. In effect, source estimation yields a spatial filter for each brain source from which to compute its activity level as a weighted sum of the measured EEG at each sensor location.

Our approach to source estimation involves the use of a finite difference model (FDM). A finite difference model captures the complex head geometry, and it accurately describes the boundary conditions of different tissues with different conductivity values (including skull orifices) (see Figure 2). The linear decomposition within this "forward" model therefore respects the physical relation between the specified components (neural sources) and the measured scalp potentials. The output from the forward modeling process leaves us with the relationship (in the format of matrix K) of how known brain activity (j) gives rise to scalp potentials (ϕ) at any given time slice:

$$\Phi = K\,j$$

The scalp potentials are given by the EEG measurements, whereas the underlying brain activity is unknown. In order to estimate the intensity of the brain activity at each cortical source location, a model that operates in the opposite direction than the forward model is required. For this so-called inverse modeling process, the projection from EEG to brain activity is captured by matrix G in:

$$j = G \cdot \phi$$

As opposed to the nonlinearity of the equivalent dipole source localization method as well as its dependency on user intervention, an advantage of linear inverse distributed source modeling is that it can be accomplished automatically. (Distributed source models are based on reconstructing brain electric activity at predefined locations in a 3-D grid of fixed solution points simultaneously.) Only amplitudes, and if necessary orientations, of sources at these solution points need to be estimated from scalp EEG, and therefore, the equations for distributed participants to categorize pictures according to those that contained animals and those that did not. These authors demonstrated that within 150 ms of stimulus onset ERPs associated with correct animal and nonanimal categorizations begin to diverge. More recent studies using stimulus presentation rates between 3 and 5 Hz show that at ~270 ms after stimulus onset, potentials recorded over occipital scalp locations differentiate negative and positive valence images from neutral images (Junghöfer et al., 2001; Schupp, 2004). We use these previous findings to guide the initial ERP analysis.

Participants

Three expert nongovernment satellite image analysts were recruited for this study. All participants had formal training as military image analysts, with an average of eight years of experience. Two were males, all were right-handed, and all had normal or corrected-to-normal vision. Participants provided consent prior to their participation and were paid $75/hour (averaging at total of $360) for their participation.

EEG Acquisition

The EEG was acquired using a 256-channel HydroCel Geodesic Sensor Net (Electrical Geodesics, Inc., Eugene, OR). All electrodes were kept below 70 KΩ (Ferree et al., 2001). All recordings were referenced to Cz. The EEG was bandpass filtered (0.1 to 100 Hz) and sampled with a 16-bit analog-to-digital converter at 250 samples/s.

Stimuli

The stimuli were broad area commercial satellite images (DigitalGlobe, Inc., Longmont, CO, 2000; 2002; 2003). On average, each broad area image covered approximately 350 square kilometers and contained either helipads, SAM sites, or AAA sites. Smaller images were obtained from the broad area images by reducing the broad area images into smaller image chips. For the chipped images that contained targets, the targets were centered. Nontarget image chips were derived from the same broad area as the target containing chips. Therefore, the nontarget chips served as "matched" nontarget stimuli. Each chip image was 500×500 pixels and covered approximately 900 square meters (see Figure 3 for example images).

Two unique sets of images were created. One served as a training set for data acquisition and the other for the test set. Within each set, there were 8,300 unique images. Within each set, 156 contained helipads targets, 24 SAM targets, and 20 AAA targets.

RSVP Paradigm

Images were presented at a frequency of 10 Hz (i.e., one every 100 ms) using a blocked design. For helipads targets, there were four blocks of eight hundred trials and one block of seven hundred trials. Within each block, 4 percent of the images contained targets. For SAM sites there were two blocks of eight hundred trials and for AAA sites there were three blocks of eight hundred trials. Within each SAM and AAA block, 1 percent of the images contained targets. Targets were randomly distributed within each block. All subjects completed the helipad blocks first, followed by the SAM and AAA blocks.

Prior to a run of each target type, all participants were provided with three samples of each target type. Participants were instructed to press a button using their index finger of either hand whenever they saw an image that contained a target of the specified type. The entire experiment, including EEG preparation and both the training and test runs, took approximately sixty minutes.

I. Introduction

For certain visual image classification (triage) applications, such as tumor diagnosis ([1]) and satellite surveillance ([2]), human experts substantially outperform computer vision systems at parsing scenes and recognizing target objects. The main challenge is that the number of qualified human visual analysts are limited and the number of visual images are enormous due to advancements in medical imaging and remote sensing techniques. The vast number of images generated can overwhelm the human analysts and promote poor decision processes ([3]).

Effective real-time image classification techniques need to be developed to reduce massive volumes of imagery in a timely manner to a much smaller subset that merits further scrutiny. (A brain-computer interface (BCI) system is one of the novel approaches that can be used to capture the cognitive signatures (unique patterns) of the brain associated with visual target detection by incorporating multiple disciplines: neurophysiology, signal processing, and classification.)

Electroencephalography (EEG) is the neurophysiologic measurement of the electrical activity of the brain from electrodes placed on the scalp ([4]). EEG-based systems have been developed to optimally couple the superior human visual processing capability with the superior signal processing and classification capability of high-speed computers ([5], [6]). Such a system ([6]) improves image-processing throughput/speed in high-volume visual classification applications ranging from medical image analysis to satellite reconnaissance.

Human EEG signals represent the aggregate activity of millions of neurons on the cortex and have high-time resolution (enabling the detection of changes in electrical activity in the brain on a millisecond-level). Evidence suggests significant amplitude differences between trial-averaged EEG responses triggered by target stimuli (event related potentials, or ERP) and trial-averaged EEG responses triggered by target stimuli ([7]). Benefits of integrating EEG responses across multiple trials include suppression of spontaneous background EEG and to enhance the event-relevant EEG saliently (in other words, averaging improves the signal-to-noise ratio).

However, in the context of accelerating image analysis done by experts, single-trial detection is necessary because repeated presentation of the same stimuli compromises system efficiency in terms of reduced processing throughput. The biggest challenge of single-trial detection is to overcome the low signal-to-noise ratio problem imposed by spontaneous background EEG responses. Spontaneous responses usually have larger amplitude than event-related responses and could completely obscure the latter ([8]).

Recent advances in adaptive signal processing have demonstrated significant single-trial detection capability by integrating EEG data spatially, across multiple channels of high-density EEG sensors ([5], [9]). A weighted sum of all electrodes over a predefined temporal window can be used as a new composite signal that serves as a discriminating component between responses to targets and distractors.

For both the trial-averaged approach and the spatially integrated single-trial approach, it is known that the event-related EEG response triggered by target detection is most prominent at a certain critical time period after stimulus onset. For example, Thorpe et al. ([7]) reported that the trial-averaged ERP generated on target versus distractor trials diverges very sharply at 150-200 milliseconds after stimulus onset for a go/no-go image categorization task. Parra et al. ([10]) achieved significant single-trial classification performance by applying the logistic regression linear classifier to EEG data in a predefined temporal window, centering around the time where the target trial-averaged ERP most sharply diverges from the distractor trial averaged ERP.

However, the brain response to visual stimuli is not a stationary pulse; instead, it reflects neurophysiological activities located in selectively distributed sites of the brain, evolving with a continuous time course. To capture the evolving spatio-temporal pattern, we propose to divide an extended ("global") EEG data epoch (e.g., 900 msec), time-locked to each image stimulus onset, into multiple nonoverlapping smaller ("local") temporal windows (e.g., 50 msec wide). Instead of applying a linear classifier on a single local temporal window, linear classifiers can be applied on multiple local temporal windows to capture their individual signatures. To capture the evolving phenomenology over the extended temporal window, outputs from multiple local classifiers can be fused to boost the overall detection performance. Our team first proposed this kind of two-level fusion scheme in 2006[1].

In this paper, we evaluate the performance of fusing the multiple local classifiers' outputs both at the feature-level and at the decision-level to discriminate between target and distractor trials (Figure 1). Two feature-level fusion methods (a Linear Discriminant Analysis Classifier known as Fisher's linear discriminant and a Relevance Vector Machine classifier known as RVM) and one decision-level fusion method (Bayesian decision fusion) are implemented and their

---

[1] DARPA Neurotechnology for Intelligence Analysts (NIA) Project PI Meeting, New Mexico, 2006.

Neurotechnology for Intelligence Analysts (NIA)

> *Exploit modern neuroscientific techniques to improve information throughput and the quality of imagery analysis*

Problem

- Computer vision efforts fail to recreate accuracy and flexibility of the human visual system.

- With a finite number of specially trained analysts, only a fraction of the enormous volume of today's imagery data collected is actually analyzed.

Solution

- Characterize neural signatures for target detection in the human brain through the following: static, broad area, and video imagery.

- Demonstrate and measure neural signatures in real-time for online classification of target detection signals in brain.

NIA seeks to demonstrate the feasibility of real-time imagery triage based on neural signatures of target detection for static satellite imagery

Neurotechnology for Intelligence Analysts (NIA)

> *Exploit modern neuroscientific techniques to improve information throughput and the quality of imagery analysis*

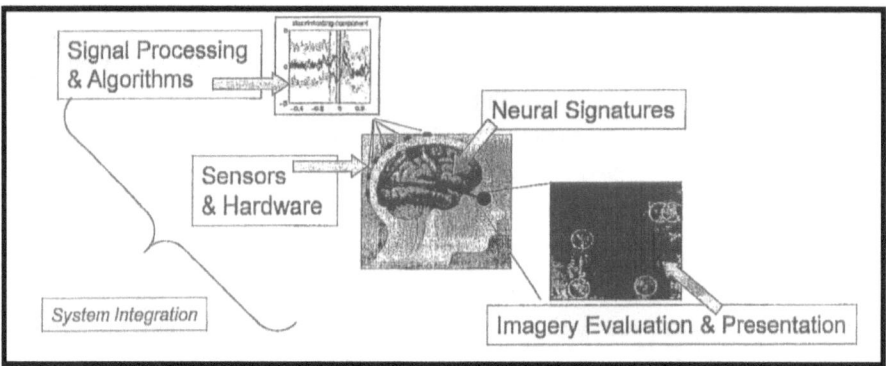

NIA goals are the following:

- Demonstrate the feasibility of imagery "triage."
- Satellite imagery is shown very rapidly to the analyst focused on target detection.
- Target-evoked neural signals are collected and classified in real-time.
- Corresponding imagery is sorted based on its neural-based classification-images of interest versus those that contain none.

DARPA-NGA Industry Day

Revolutionary Imagery Analysis through Neurocognitive Techniques

Dr. ???
DARPA/DSO???
???se@darpa???

Why Neuroscience?

- Analysis is a complex cognitive activity
- Analysis is done in the context of ancillary/collateral information

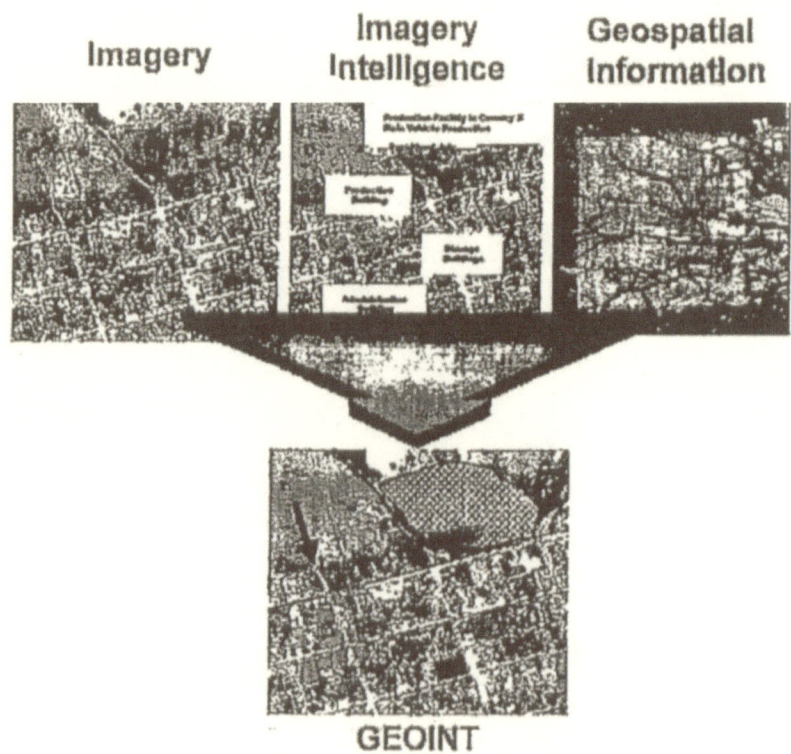

Imagery
Imagery Intelligence
Geospatial Information

GEOINT

Classification of target images was .79, which also reflects substantial improvement over classification based on the behavioral data.

Discussion

In this paper, we present a single-trial analytic framework for EEG data. The results from our example study, using this framework, show that classification of satellite images using neural signatures of target recognition can be accomplished. The performance statistics show that the classification is significantly above chance levels. In the case of classification of target images, the performance is remarkable in that classification based on participants' neural response was superior to classification based on participants' behavioral responses. Another remarkable aspect of the classification performance was that the discriminant weights derived from the training set were successfully applied to a new data set.

For nontarget images, the classification was close to chance levels. One relevant factor was the relatively low resolution of these commercial images. However, in the break between RSVP runs, all participants reported that they saw "target-like" objects, in nontarget images that were of significance to image analysts. Based on the subjects' reports and the classification performance of the nontarget images, we randomly selected eight nontarget images that the analysts classified as targets. An image analyst who did not participate in the study performed a poststudy review of these images and found that seven of the images contained intelligence-related content. Although these features were not designated as targets, image analysts attended to them nonetheless, perhaps illustrating the automated nature of domain-specific processing by experts. Clearly, if the present neurotechnological procedure were to be applied to practical target detection in satellite images, a second round of inspection would be required to separate possible targets of military interest (detected automatically by the analysts brains) from those of interest to the particular intelligence mission.

The analyses presented here rely on the assumption that the interesting information in the brain's response is captured by time—and phase-locked activity. This is because the identification of the relevant brain regions and their time courses was based on the ERP. In addition to being time—and phase-locked, there is an additional assumption that the activity is stationary (i.e., does not vary substantially in latency from trial to trial). These assumptions are consistent with much of the ERP literature. For example, two ERP components that differentiate between target and nontarget images are the N270 and P300. The N270 has been shown to differentiate images that are highly arousing from those that are neutral when presented in an RSVP paradigm (Junghöfer et al., 2001; Schupp, 2004). These previous findings suggest that individuals automatically attend to highly arousing stimuli and the present findings suggest that targets, when attended to, may engage similar arousal mechanisms. The P300, from previous research, has been related to memory processes (Farwell and Donchin, 1988), and in the present study, likely reflects memory processes associated with target detection.

However, it is now well-known that important brain activity in response to cognitive and perceptual events is not time—and phase-locked. These "induced" EEG responses (Tallon-Baudry et al., 1996; Luu et al., 2004) represent frequency-specific changes in the EEG and are best detected on single trials. Moreover, cognitive functions result from the coordination of multiple brain regions (i.e., involve a network of brain structures). Our framework does not rule out analytic procedures that are sensitive to nonstationary signals or cortical coupling.

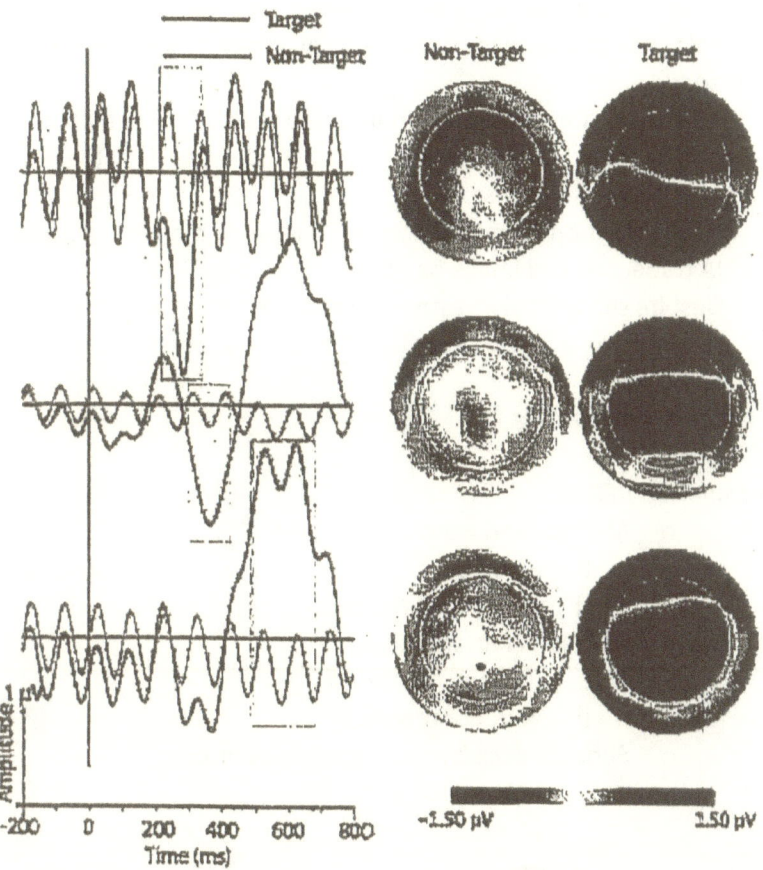

Figure 4 Grand-average ERP waveforms of participants at three recording sites.

 # Problem

- Imagery analysts are currently faced with an enormous volume of data, which is growing exponentially as new collection platforms are added.
- Analysis is done by a finite number of specially trained analysts, and only a small fraction of imagery data collected is actually analyzed with today's systems.
- New collection platforms coming online will mandate a revolution in IA tradecraft and image handling.
- Despite substantial investment, computer vision efforts have failed to recreate the accuracy and flexibility of the human visual system.

NGA Imagery Analysis

- Exploitation Focus
  - Monitor and watch critical issues
    - warning
    - sustain awareness
  - Detailed analysis of hard targets
    - find and monitor
    - in-depth research
  - Support decision making and operations
    - tailored support

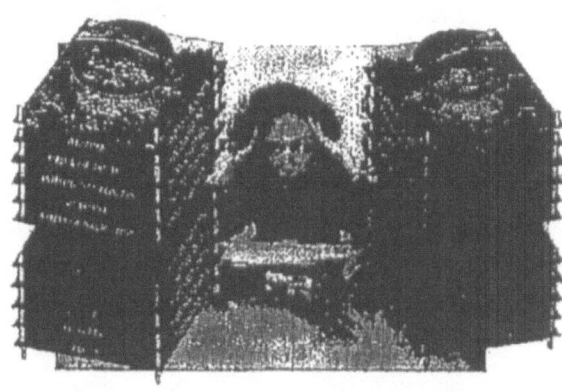

Challenges

- Accumulated knowledge and experience within the IA community is often lost

    o   Is there a way to capture, manage, and retain these critical details?

- IA tradecraft has not been systematically codified

    o   Is there a way to understand the analytic process that gets beyond subjective measures alone?

- An "IA-centric" approach to the design of systems is badly needed

    o   The IA is human capital of NGA—how can we enable them to do their job more effectively?

☐ Goal

- Efforts should:

  - ○ Assist IAs with reaching NEW analytic insights

    - ☐ *Rapid insights with increased confidence*

  - ○ Create novel tools for DISCOVERY of new information and patterns

    - ☐ *Tools that help synthesize, not just create a more complicated interface*

    - ☐ *Support the new forms of imagery now available (nonliteral)*

☐ First Steps

- Understanding the analytic process

    o We now have tools and techniques from neuroscience, cognitive science, human factors, and psychology that can enable a greater understanding of these processes

- Add metrics to a formerly subjective process

    o We hypothesize that like other complex decision making tasks, there are fundamental cognitive building blocks that enable analytical insights and pattern recognition

- If we can uncover these fundamental pieces, we can begin to train to these specific cognitive skills

Operational Neuroscience!

☐ **The Continuum of Operational Neuroscience . . . .**

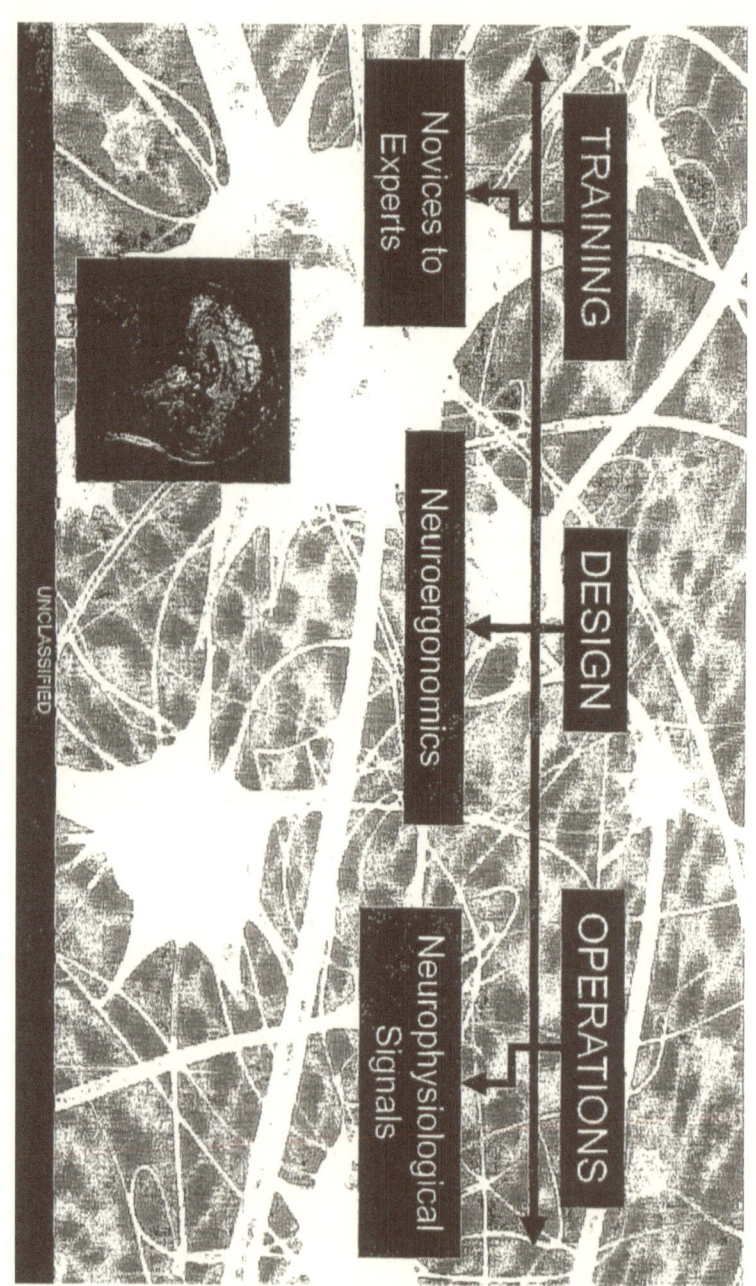

42

☐ Training: Novice to Expert

Brain differences during skill development

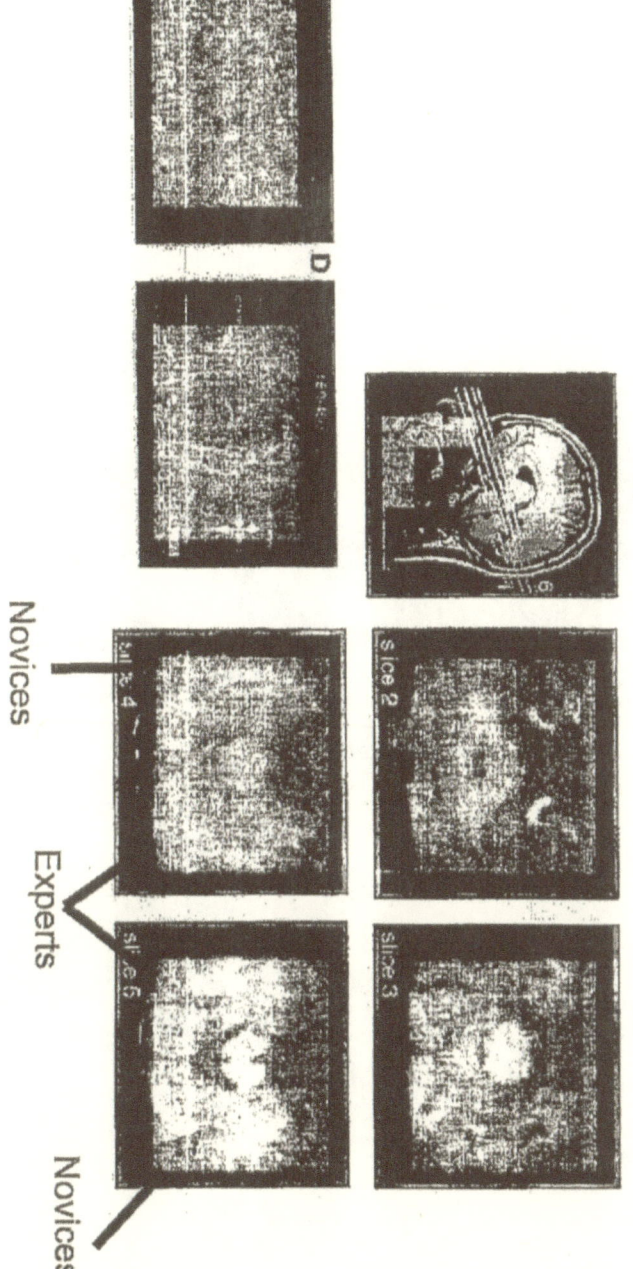

Accelerated Learning

Question No.1: Can we use neurotechnology to improve learning?

- Answer: Yes!
- Using a combination of three technologies:

    1. Image brain activity during learning
    2. Identify brain networks that change with learning
    3. Stimulate these brain networks to accelerate the learning process

- Using this method, we have found over 100 percent improvement in learning for the same amount of training

Program Vision

- Utilize neuroscience to understand the development of expertise through learning
- Use this understanding to directly facilitate and *accelerate* task learning for the warfighter

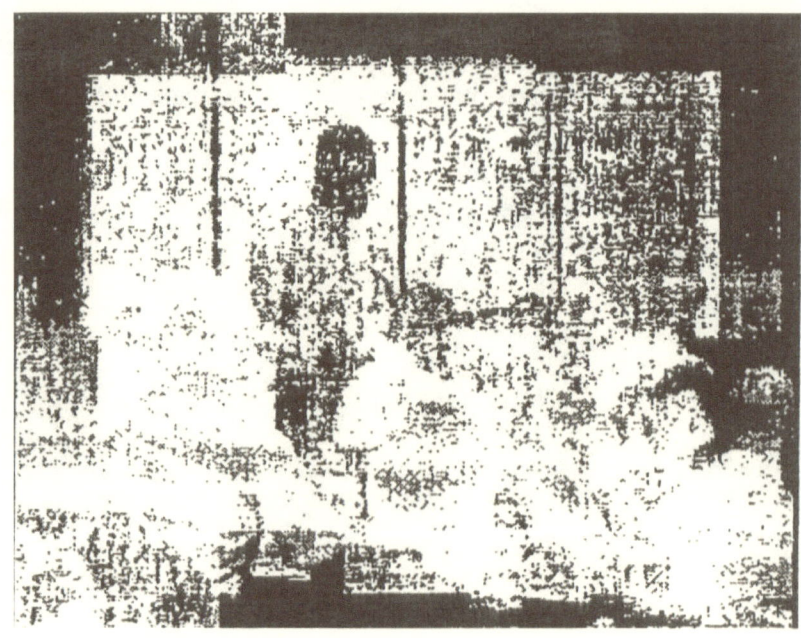

- Learning is a continuous challenge in the operational environment
- Current methods of learning fail to capitalize on basic lessons from neuroscience
- Measures of learning on key skills only as good as *qualitative and subjective* assessments

Using a neuroscience-based approach, change the paradigm of learning in the military

Phase I—Goals

Utilize neuroscience to understand the development of expertise through learning to directly facilitate and accelerate task learning for the warfighter.

Program Goals

- Identify the neural basis of expert performance
- Track progression from novice to expert with classification of intermediary states
- Demonstrate a two-fold increase in the progression between stages of the novice-to-expert path

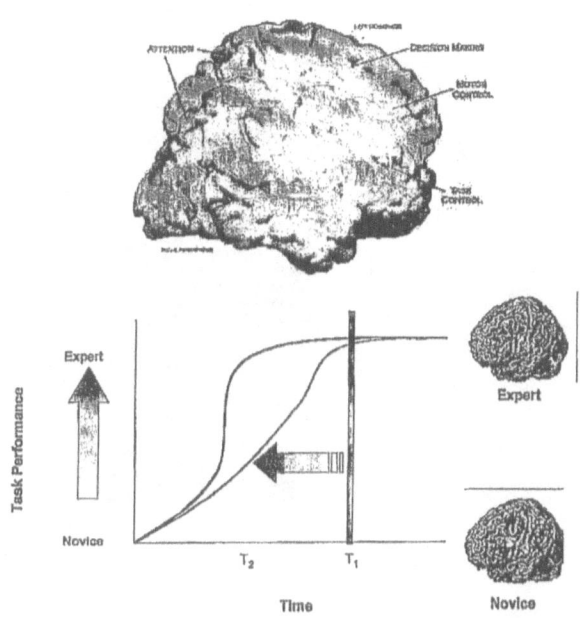

Accomplishment—
Remote view Integration

Triage integration approach maintains context before RSVP and during verification

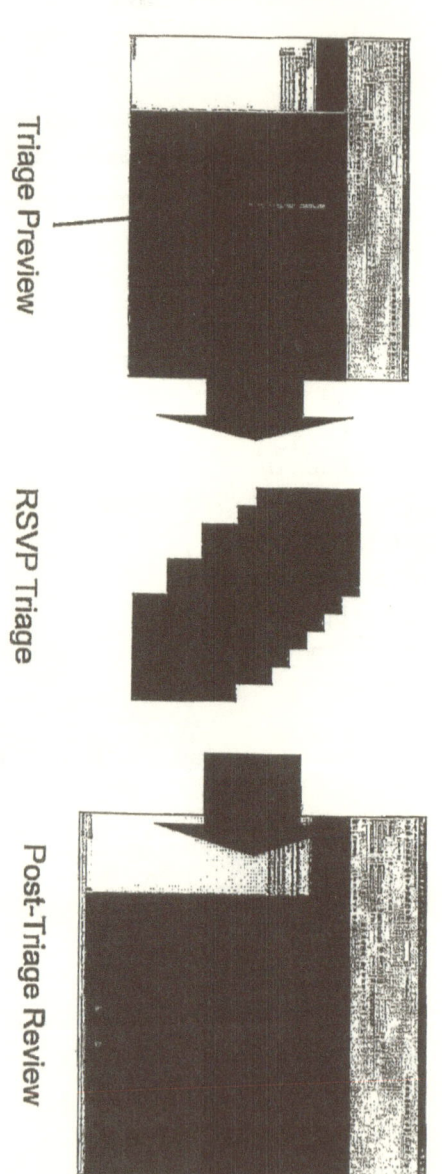

Triage Preview

RSVP Triage

Post-Triage Review

Analysts maintain "global view" before triage begins and can "fly to" any target hit results during posttriage exploitation

Accomplishment—
Uncovering impact of image complexity

As targets become less stereotypical, the image stimulus duration necessary for correct detection increases.

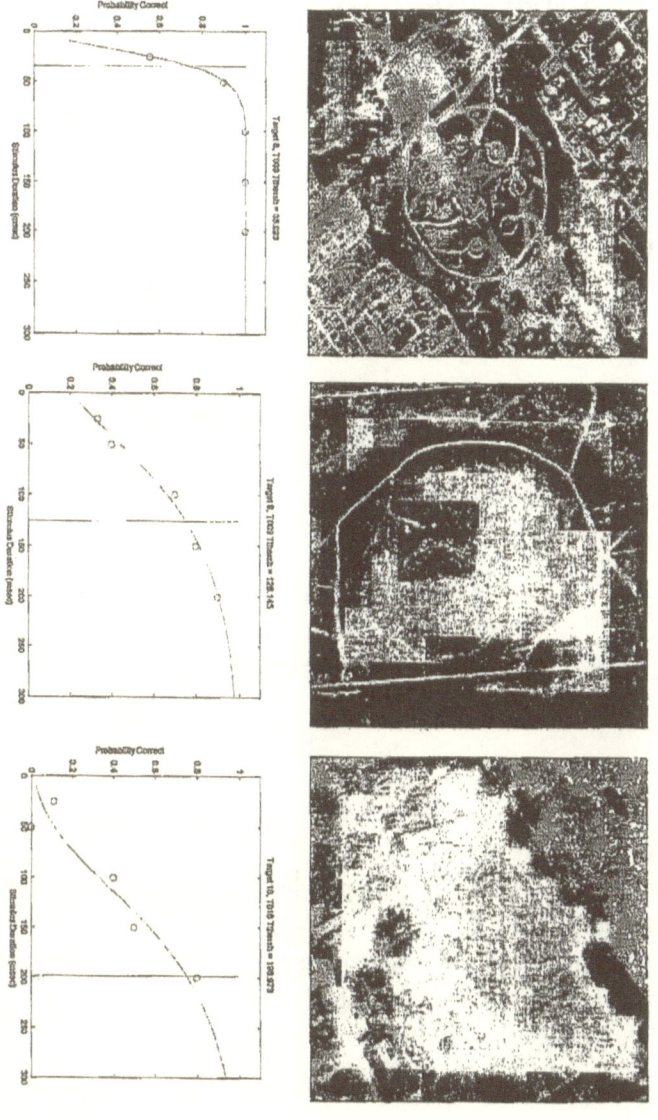

Accomplishment—
Computer vision enables target centering, enhancing detection

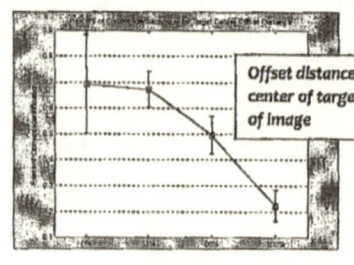

Offset distance from center of target to center of image

Human detection performance is weaker when targets are off-center

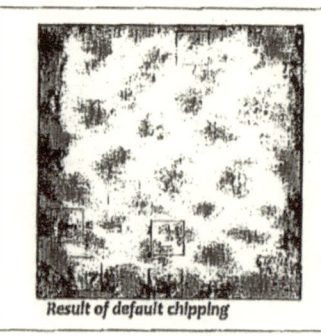

Result of default chipping

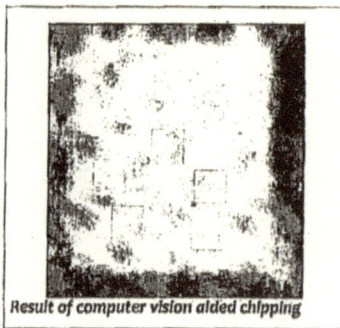

Result of computer vision aided chipping

- Computer vision techniques identify objects that share features with target of interest
- Small adjustments are made so that chipped images place likely targets at center of each chip
- Approach employed simple features (edges, intensity, spatial filter responses, etc.) that are easy to implement and relevant across target types
- All targets in validation set were better centered

Phase I—Approaches

Optimized Task Qualities	Neurofeedback	Global Network Measures	Direct Stimulation
• Apply multiple biological and cognitive findings about the user to customize the learning environment	• Present the user with real-time feedback on brainwave activity in the form of a haptic feedback during training	• Determine and facilitate the neural mechanisms of consolidation (declarative and procedural memory, attention networks, and memory chunking)	• Utilize techniques such as tDCS in combination with functional imaging to directly stimulate neural pathways critical for learning

Accomplishment—
fMRI Signatures of Novice vs. Expert Behavior

Threat-Nonthreat Stimuli

Novice

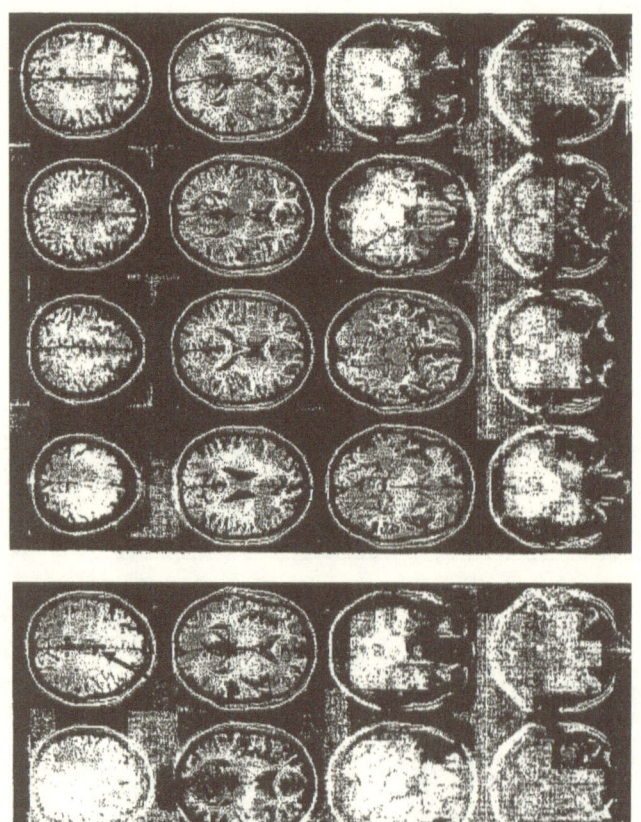

Expert

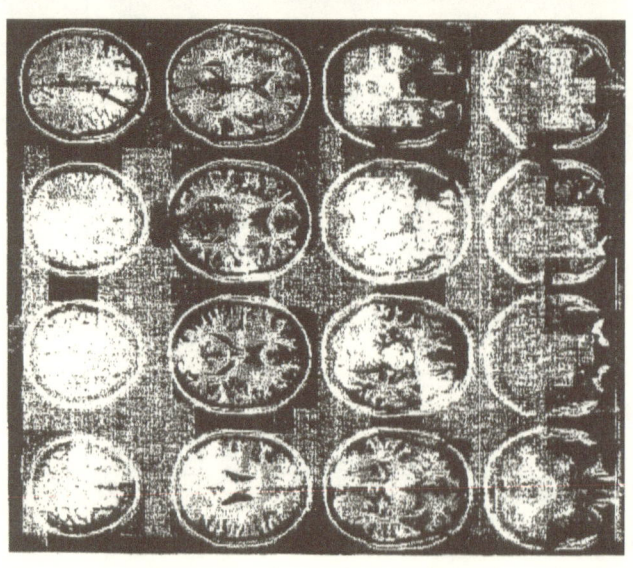

52

Accomplishment—Direct Stimulation leads to 2 × Improvement

tDCS stimulation applied to the right sphenoid (right temple) at two milliamps for thirty minutes provides an improvement in learning vs. sham in threat detection training

- 2.1 × improvement (p = 0.0093) in threat and nonthreat detection accuracy

- 3.1 × improvement for threats alone (p = 0.0004)

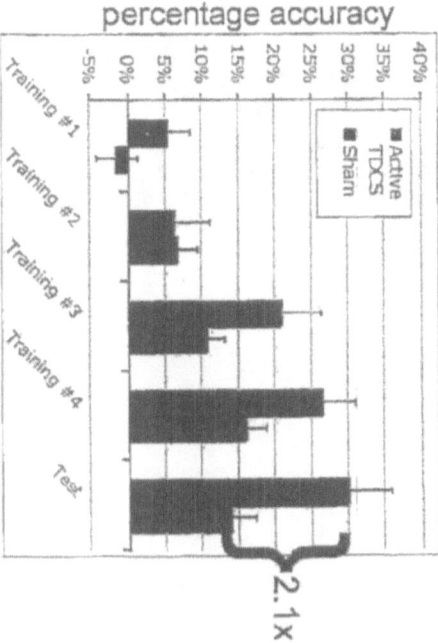

Phase I—Results

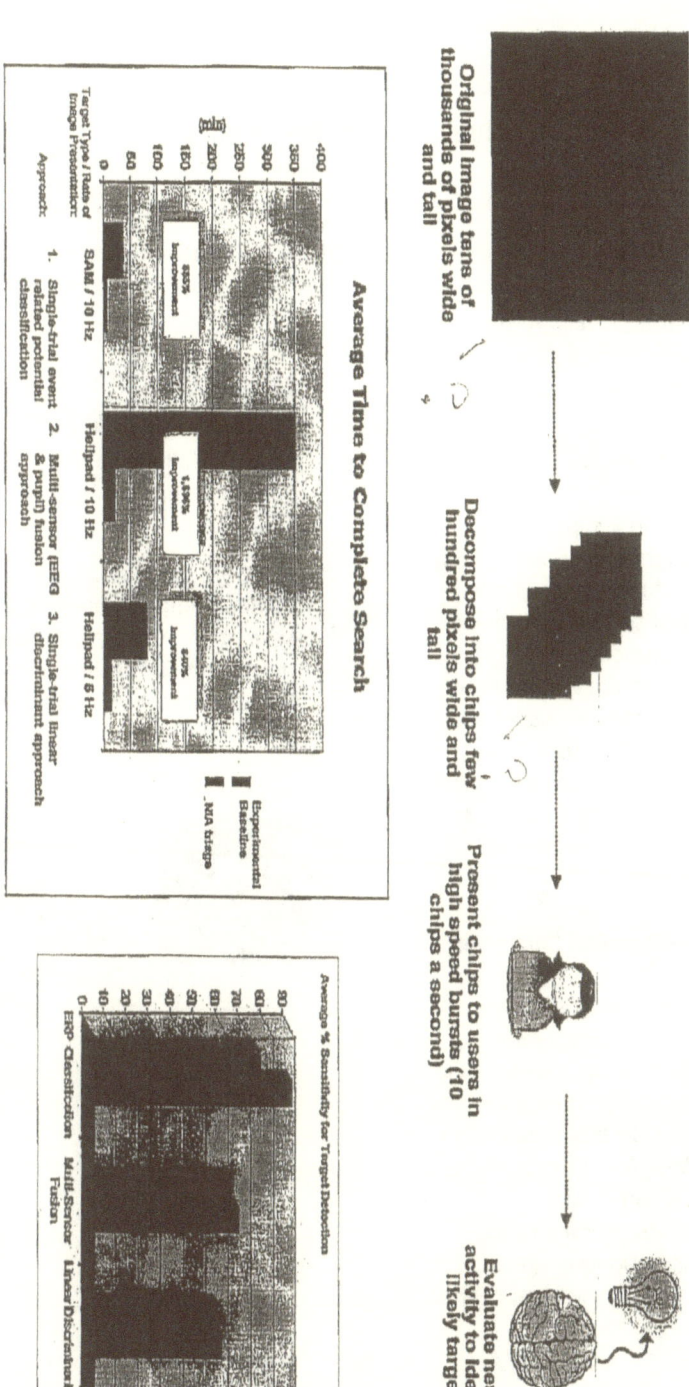

Original Image tens of thousands of pixels wide and tall

Decompose into chips few hundred pixels wide and tall

Present chips to users in high speed bursts ('10 chips a second)

Evaluate neural activity to identify likely targets

Average Time to Complete Search

Target Type / Rate of Image Presentation:

Approach:

1. Single-trial event related potential classification
2. Multi-sensor (EEG & pupil) fusion approach
3. Single-trial linear discriminant approach

Experimental Baseline
N/A times

SAM / 10 Hz — 893% Improvement
Helipad / 10 Hz — 1.69% Improvement
Helipad / 5 Hz — 540% Improvement

Average % Sensitivity for Target Detection

ERP Classification   Multi-Sensor Fusion   Linear Discriminant

Result: >300 percent throughput improvement and detection equal or better than current SOA

55

NIA Phase 2—Vision

Phase 2—Goal

Integrate modern neuroscientific techniques into imagery analysis workflow to improve throughput and quality of imagery analysis

Phase 2—Metrics

- Maintain 300 percent throughput increase in imagery exploitation in a realistic analyst software/hardware environment.
- Demonstrate greater than or equal to unassisted image analyst sensitivity.
- Maintain performance across three complex target classes and under variable operating conditions.

Phase 2—Technical Challenges

- Capture brain signals in real-time during realistic imagery analysis on baseline imagery exploitation systems
- Categorize target detection brain signals based on object/ scene complexity
- Integrate neuromorphic computational image analysis and physiological brain signals

Phase 2—Applied Science

Apply Phase 1 breakthrough science in operational contexts

- Extend capture of brain signals for target detection to:
    - Multiple imagery types
    - Diverse target and scene complexity
- Integrate brain-assisted search into standard imagery analysis software
- Leverage/converge with automated machine vision technologies
- Demonstrate with trained analysts with realistic tasks and environment

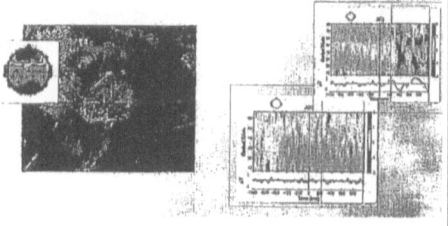

EO Dataset Image Example

Military Facility

Naval OB

POL Storage

Cargo Ships

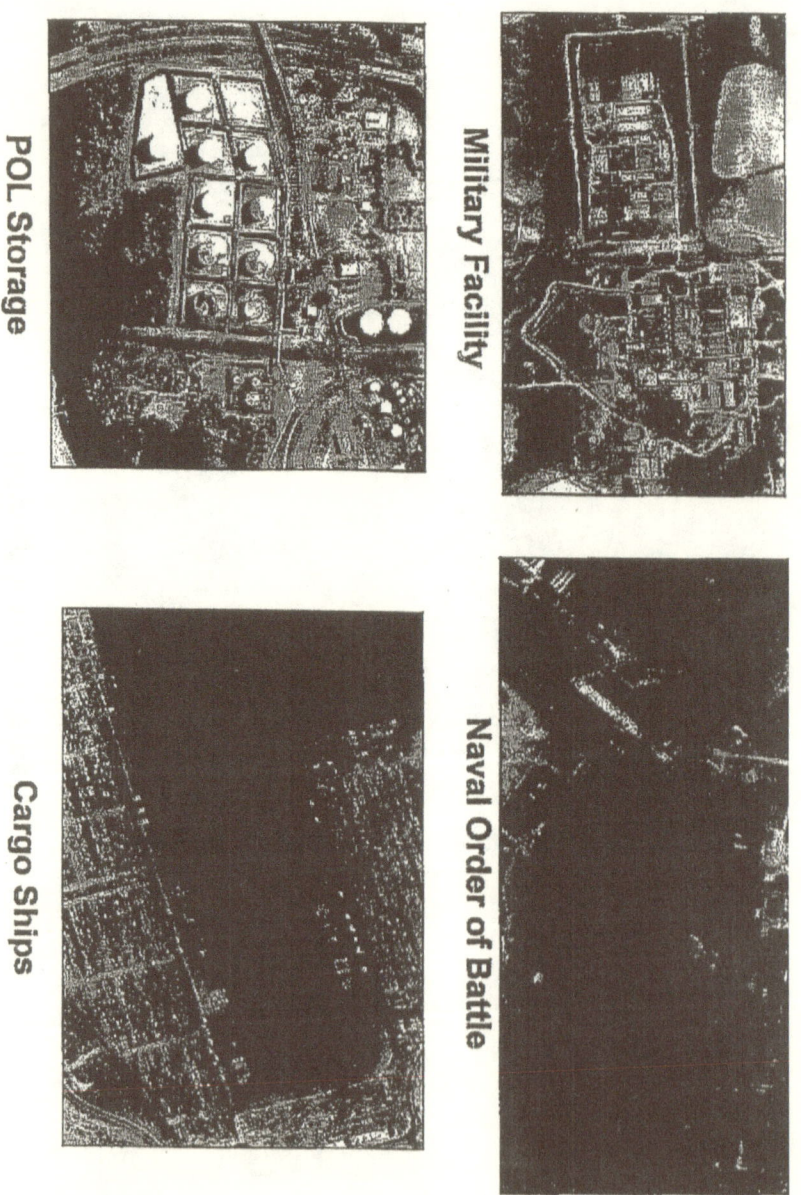

Examples of New Target Types

POL Storage

Military Facility

Cargo Ships

Naval Order of Battle

Stimulating Brain Science—
The Future of Neurotechnology

The Mind Research Network

Threat Detection Training

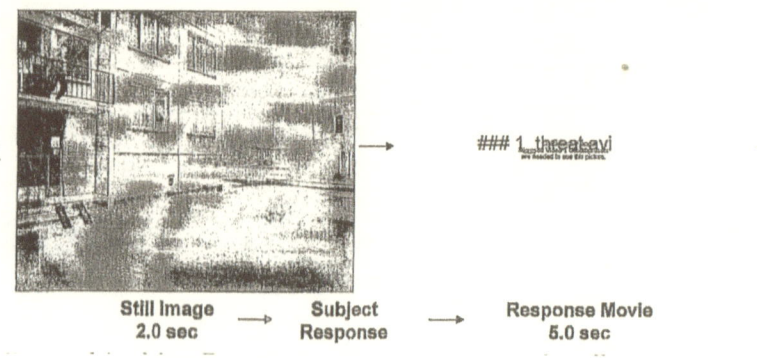

Still Image   →   Subject   →   Response Movie
2.0 sec   Response   5.0 sec

- Stimuli provided by doctors and colleagues at SNL
    - Adapted from DARWARS-NK
- Subjects indicate if threat cue is present in still image
- Response movie provides feedback
    - Threat plays out if missed
    - Four, fifteen-minute training sessions in series
    - First thirty minutes using tDCS

Baseline Testing

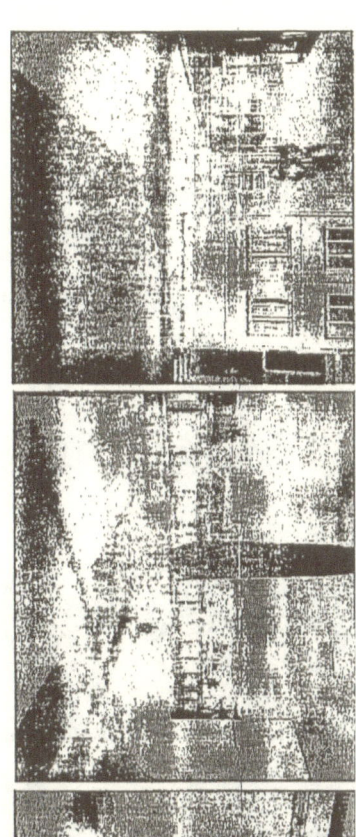

- Subjects perform baseline testing before and after training
- Increase in accuracy from pretraining baseline to posttraining used to quantify learning
- Difference in learning between groups receiving active and sham tDCS reveals effects on learning

Measures of the Novice and Expert States

Threat-Nonthreat Stimuli

Novice

Expert

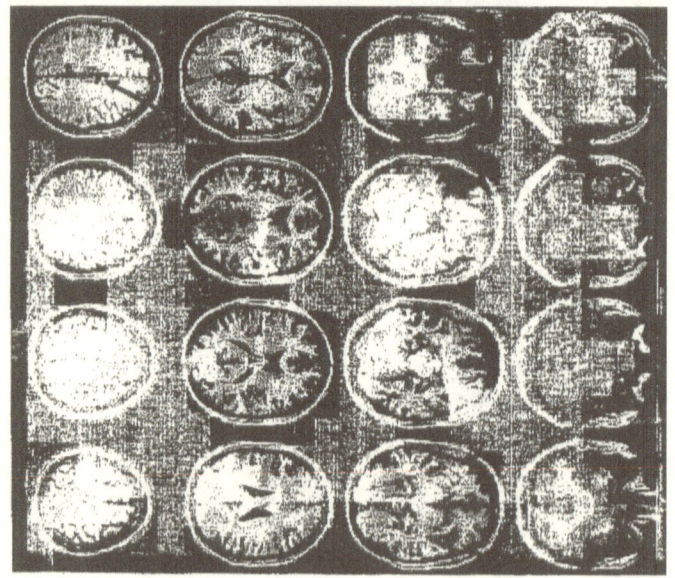

Neurotechnology for Intelligence Analysts (NIA)

—Three teams participated in NIA Phase 2

Teledyne Scientific and Imaging, LLC		
Teledyne's Phase 2 NIA system utilized an eye tracker that monitored the analyst's eye movements and gaze fixations during the viewing of imagery segments, presented at a rate of 1-2 sec/image.	The system was calibrated to individual user's brain activity by having each user search for "T"s. The resulting target-detection brain patterns were used to detect a variety of targets in multiple imagery types.	The analyst's EEG signals were time-locked to gaze fixations and used to determine probability of target detection.
 	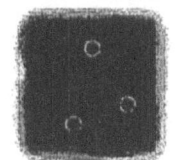	

Columbia University		
Based on a set of known targets, Columbia's computer vision algorithms detected potential targets and centered imagery segments around each potential target.	As the analyst viewed rapidly presented image segments (5-10/sec.), the system identified probable targets based on the analyst's brain signals.	The system allowed the analyst to "jump" to regions of the imagery most likely to contain targets, based on the analyst's brain signals during previous viewing of the imagery segments.
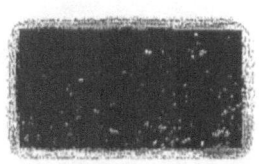		

Honeywell International, Inc.

Honeywell's Phase 2 NIA system included computer vision algorithms that automatically detected and centered potential targets within imagery segments.

The system identified targets by fusing data from the analyst's brain responses and button presses as the analyst viewed rapidly presented image segments (3-10/sec.).

Results were displayed as target probability maps overlaid on the imagery, which the analyst used to verify targets.

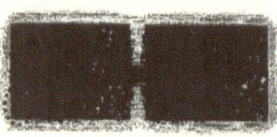 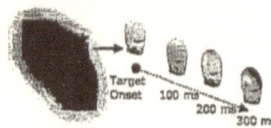 

???roscience and Eye-tracking for Rapid Image Triage

"achieving matched search accuracy in one third of the time or better!"

??? racking signals are captured as an IA ???—viewing presentation of imagery.

???archers have made significant ???he development of automated ???nition, there is still a signifi??? ???etween the performance of ???s and the capabilities of hu???.

???nonapproach engineers take in ???enting target recognition systems ???esign of task-specific image proc??? algorithms which operate on the ??? data but do not exploit task or ???-based information.

??? other hand, IAs perform extremely ??? target detection tasks due to their ???ense of task constraints and their ??? on contextual information pro??? the characteristics of the scene ??? dynamics.

Our goal—capitalize on IAs capabilities through appropriate human-machine interfaces that maximize overall performance along both accuracy and throughput dimensions by the following:

Measuring Brain Responses to Objects of Interest

Building upon overwhelming evidence in Phase I and II, we exploit neural signals, with the following features:
- Universal response—not target-specific
- Repeatable signal
- Robust temporal characteristics

Tracking eyes and focus of attention
- Using commercial, off-the-shelf desktop eye-tracking apparatus, we continuously monitor eye movements
- Permits free viewing
- Reliable as an indicator of attention and intent

Integrated with remote view
- Image search and confirmation embedded in remote view environment
- Use of familiar analyst too

EEG signals (left) and eye movement patterns (right) are monitored in real-time to classify IA's responses to imagery and determine the presence of targets.

Current capabilities deliver >3 × image throughput!

Brain-in-the-Loop Workstation Concept

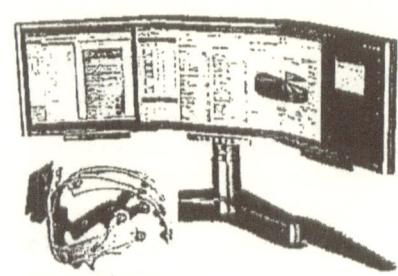

Advances in human-machine integration and ease-of-use can lead to rapid and broad deployment to permit analysis of workflow impact and new opportunities.

The increase in analyst throughput is a particularly critical objective since IAs are currently faced with an enormous volume of imagery, only a fraction of which can be processed or reviewed in a timely operational manner.

We are demonstrating a viable role for the use of brain responses and eye movements to greatly improve image analysis.

Our work is also demonstrating that signal-processing methods can be developed to permit reliable detection in the context of operational information analysis.

Brain signal monitors and eye trackers continue along a path of miniaturization and reduced complexity that will facilitate future integration.

Fully Integrated System Goal

Seamlessly integrate neuroscience with familiar analyst tools
- Compact brain cap
- Monitor-integrated eye tracker
- Brain-in-the-loop image review

Unprecedented operational impact

- Performance improvements as a function of continued use of the system
- Multiple analysts share and collaborate
- Training and mentoring

A critical priority for the successful operational deployment of these techniques is the design of optimal coupling between the human user (IAs) and the computer-based analysis subsystem.

We will conduct analysis of IA workflow to achieve a thorough understanding of this integration to achieve effective:

Brain-Driven Imagery Exploitation

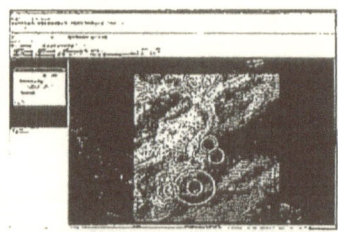

Real-time visual feedback would support improved interaction and user confidence.

Operational System Concept

ROMAN vs. NSA No. 09-2947

In this case, the NSA refuses to respond to my FOIA request, violating FOIA guidelines. I asked for records on implanting thoughts that could cause behavior. I am waiting for a decision from the court.

US District Court
Eastern District of NY

Gilbert Roman, Pro Se Plaintiff,   MEMORANDUM OF LAW

V   cv-09-2947

NSA

Under the FOIA 5 USC Sec. 552 records must be made for public review, upon request.

Under the 1 st and 14th Amendments I am entitled to free press and speech.

Under Article III injury in fact, because of all the lies, not responding, refusing to respond, denying me public records, an injury is happening.

Under Board of Ed, Island trees union school district no. 26 v pico 457 US 853 1982, Hazle Sch Dist v Kuhlmeler 484 US 260 1988 I am entitled to exercise free speech and press.

Under the Dept. of Justice Mar 2007 guidelines an agency may make discretionary disclosure.

Under ACLU v NSA 493 F3d 6646th Cir I am entitled to records.

Under Allen v Wright 468 US 737 1984 I am entitled to records.

Under the Supreme Court, has found numerous types of injuries to be suffient for standing purposes in common law, constitutional, statutory, economic, and environmental see Friends of the Earth 528 US at 182-185.

I swear the forgoing is true and correct under penalty of perjury.

Central Islip, NY

Gilbert Roman

6 27-10

US DISTRICT COURT
EASTERN DISTRICT OF NY

Gilbert Roman, Pro Se

v.

NSA

PLAINTIFF STATEMENTS OF
UNDISPUTED FACTS

CV-09-2947

I have made two different requests to the NSA.

1. I requested information on FMRI technology; which reads a persons thoughts.
2. Information on the technology that implants thoughts and may cause behavior.

This case was brought because the NSA fails to respond to request on the technology that implants thoughts and may cause behavior. The NSA refusal to process my request violates FOIA guidelines. I believe in my free country and I love it. I do not trust man and his nature to do evil and corrupt things. This is why we have a system of checks and balances, along with the FOIA.

I swear the forgoing is true and correct under penalty of perjury.

Gilbert Roman 6/27/10

Roman vs. CIA No. 09-3344

In this case I ask for my FOIA processing forms. The CIA Fail to produce any search time forms. Which means, how could search be made and no search time entered. They release several forms; one of which, is not clear. They refuse to enhance this form. This form comes from them and they have the technology to enhance. What are they trying to hide?

I present pages that show the CIA has lied and caused obstructed justice. I mentioned the book "Spies Lies and Whistleblowers." In this book, two ex-British agents stated that illegal activities are taking place in the intelligence community. That because of Nixon and Watergate, Iran-Contra affair, 44+ politicians and Rabbi's arrested in NY and NJ for money laundering and bribery in 2009; we must check. That anybody at any time in life can break the law or make a mistake.

MEMORANDUM OF LAW WAS ENTERED SITING:

1. FOIA records must be produced for public review, upon request.
2. Under the Board of Education, Island trees union school district no 26 v Pico, 457 US 853 1982, Hazle Sch Dist. v Kuhmeler, 484 US 260 1988 I am entitled to free speech and press.
3. Under Nazi war crimes act public law 105-246, I am entitled to records.
4. Under the classified information act sec. 2 at any time a pre-trial conference can be held to review classified material.
5. Under the Dept. of Justice guidelines Mar. 2007 an agency may make discretionary disclosure.
6. That I believe because of all of these things that an Article III issues is taking place, Injury in Fact.
7.

Justice for all, this is what I believe. Yet not seen yet in my cases. I present the un-clear document for your review on the next page.

05390973

D/t (b)(3)

PRIVACY ACT REQUEST

SUBJECT: (Optional)	RE: INFO ON SELF				REQUEST NUMBER	
	ROMAN, GILBERT				P96-1518	

FROM:		EXTENSION	DATE SENT	5 August 1996
IP&CRD/MSG/OIT			SUSPENSE DATE	

TO: (Officer designation, room number, and building)	DATE		OFFICER'S INITIALS	COMMENTS (Number each comment to show from whom to whom. Draw a line across column after each comment.)
	RECEIVED	FORWARDED		
1. DDA/IRO-Mr.Hatch 1236 Ames Bldg.				Please process the attached Privacy request.
2.				
3.				PLEASE PROCESS UNDER PA & FOIA
4. RETURN TO: DDA/IRO-Mr.Hatch 1236 Ames Bldg.				SEARCH CUT OFF DATE: 30 July 96
				FEE CATEGORY: No Fees
5.				
6. DDO/	13 Aug. 96			PLS NOTE: CR: P96-8018
7.				*NOT clear why other Documents are clear*
8.				Subject has been searched with all ...counts of this Directorate which ...have an interest or record. This ...ctorate has no information which ...be ident ...
9.				
10.				Directorate of Operations
11.				ACTION: DDA/IRO, DDO/IRO
				INFO: DDA/IRO
12.				APPROVED FOR RELEASED DATE: 10-Jun-2009

RETURN TO:
IP&CRD/MSG/OIT 1107 Ames Bldg.

PRIVACY ACT REQUEST

FORM 3834 OBSOLETE PREVIOUS 12-76 EDITIONS

NO time entered for searches

ROMAN vs. DARPA No. 09-5633

I wrote twice to DARPA asking for FMRI technology records. They refused to respond. Only after starting this case did they agree to respond. In fact, they are the agency that released the 183 documents on FMRI technology. Some of these pages are in this book. Waiting for the court to order further evidence released (date of development, first report on first person used against successfully, FMRI technology)

ROMAN vs. NSA, CIA, NRO No. 09-4281

I ask for records from each agency. All refused to answer or process my request properly. I make the same and similar memorandum of law entries in this case.

Woe, to the country who denies freedom and liberty.

Woe, to the country who denies due-process, free speech you see.

Woe, to my country, who I love in my deeds.

I am waiting for the courts answer also.

ROMAN vs. NRO No. 09-2504

In this case, the NRO misplaced my first request, refused to respond and denied having any records on FMRI technology. The claim on not having records on FMRI technology is proven false in the 183 documents released by DARPA. Clearly, in these DARPA exhibits, DARPA mentions satellites, FMRI, retrieving data. The NRO designs, builds and operates reconnaissance satellites for the Dept. of Defense and intelligence community. These lies were presented to the court. I also ask for FOIA processing forms on these request. Letter dated November 2, 2009, states that the NRO is searching for the records. We are in November 2010, maybe searches were never made. I entered similar memorandum of law in this case also. I request discovery, hearings and court orders to produce records. Waiting for the NRO and courts decision.

Now Nov 2010 and still
waiting for an answer

2 November 2009

Gilbert Roman

Remember the book Spies, lies and
whistleblowers. That if a proper two
fold entre is not done a no record
response will be given (pages 115-116)

Dear Mr. Roman:

This is in response to your letter dated 25 October 2009, received in the Information Management Services Center of the National Reconnaissance Office (NRO) on 30 October 2009. Pursuant to the Freedom of Information Act (FOIA), you are requesting:

1) ...."copies of all the Freedom of Information Act and/or Privacy Act of 1974 task sheets used to process my request to your agency; which you responded on July 1, 2009 and Oct. 15, 2009....."

2) "Copies of the DUTY OFFICERS forms authorizing these searches to be made on July 1, 2009 and Oct. 15, 2009; Your response letters to my request for FMRI technology."

3) "Copies of the forms from the CLASSIFIED DOCUMENT RECEIPTS authorizing searches for my request for FRMI technology; which you responded on July 1, 2009 and Oct. 15, 2009."

4) "Copies of the forms from the OFFICE OF THE CLASSIFIED DOCUMENT REGISTER OF CONTROL Authorizing searches my request for FMRI technology; which you responded July 1, 2009 And Oct. 15, 2009."

5) "Copies of any and all memorandums, e-mails concerning Gilbert Roman (ME)."

We have accepted your request. It is being processed in accordance with the FOIA, 5 U.S.C. § 552, as amended, and the NRO Operational File Exemption, 50 U.S.C. § 432a. Unless you object, we will limit our search to NRO-originated records existing through the date of this acceptance letter.

Since we may not respond within the 20 working days stipulated by the Act, you have the right to consider this as a denial and may appeal to the NRO Appeal Review Panel. It would seem more reasonable, however, to have us continue processing your request and respond as soon as we can. You may appeal any denial of records at that time. Unless we hear from you otherwise, we will assume that you agree, and will proceed on this basis.

The FOIA authorizes federal agencies to assess fees for record services. Based upon the information provided, you have been placed in the "other" category of requesters, which means you are responsible for the cost of search time exceeding two hours ($44.00/hour) and reproduction fees (.15 per page) exceeding 100 pages. Additional information about fees can be found on our website at www.nro.gov.

In your request you expressed a willingness to pay "after clearance by me." We will notify you if it appears that assessable fees will meet or exceed our $25.00 minimum billing threshold.

Regarding your request for a fee waiver, please be advised that fee waivers or reductions are granted when there is a public interest in disclosure of information, which will contribute significantly to the public's understanding of the operations or activities of the NRO. A decision to waive or reduce fees cannot be made until after any responsive documents to be disclosed have been reviewed for "public interest".

You have the right to appeal this determination by addressing your appeal to the NRO Appeal Authority, 14675 Lee Road, Chantilly, VA 20151-1715 within 60 days of the date of this letter. Should you decide to do so, please explain the basis of your appeal.

If you have any questions, please call the Requester Service Center at (703) 227-9326 and reference case number F10-0034.

Sincerely,

Stephen R. Glenn
Chief, Information Access
and Release Team

NATIONAL RECONNAISSANCE OFFICE
14675 Lee Road
Chantilly, VA 20151-1715

16 June 2009

Gilbert Roman

Dear Mr. Roman:

This is in response to your letter, dated 14 May 2009, received in the Information Management Services Center of the National Reconnaissance Office (NRO) on 20 May 2009, and your subsequent correspondence, dated 27 May 2009. In your 14 May letter, pursuant to the Freedom of Information Act (FOIA), you requested:

"1. ...information on functional magnetic resonance
    imaging.
 2. The date it was put into service.
 3. The first successful report on the first person it was
    used on successfully."

We have accepted your request. It will be processed in accordance with the FOIA, 5 U.S.C. § 552, as amended, and the NRO Operational File Exemption, 50 U.S.C. § 432a. Unless you object, we will limit our search to NRO-originated records existing through the date of this acceptance letter.

Since we may not respond within the 20 working days stipulated by the Act, you have the right to consider this as a denial and may appeal to the NRO Appeal Review Panel. It would seem more reasonable, however, to have us continue processing your request and respond as soon as we can. You may appeal any denial of records at that time. Unless we hear from you otherwise, we will assume that you agree, and will proceed on this basis. Fees are incurred whether we find responsive records or not.

The FOIA authorizes federal agencies to assess fees for record services. Based upon the information provided, you have been placed in the "other" category of requesters, which means you are responsible for the cost of search time exceeding two hours ($44.00/hour) and reproduction fees (.15 per page) exceeding 100 pages. We will notify you if it appears that assessable fees will exceed our $25.00 minimum billing threshold.

As mentioned previously in our response letter to you, the law requires that requesters express a willingness to pay fees for this service. We have not yet received that willingness to pay from you. As a courtesy to you, in this case, we will proceed with the search up until the point that fees are incurred. You will receive the benefit of two hours of search time and 100 pages of copying. At that point, we will notify you that we have begun to incur costs and ask that you provide us with that willingness to pay fees before we continue. You may specify a limit at that time.

With regard to the issue of your 3 March 2009 request, we have been able to track the receipt of your correspondence within the NRO's Mail Processing Center on 9 March 2009. It appears, however, that the correspondence was lost or misdirected internally, and did not arrive at the proper destination for acknowledgement and processing. We apologize for the delay and inconvenience caused by this error.

If you have any questions, please call the Requester Service Center at (703) 227-9326 and reference case number F09-0063.

Sincerely,

Linda S. Hathaway
Chief, Information Access
and Release Team

A Civil Commission

A civil commission is needed to overcome the politicians and special-interest groups. The people's needs must be met right away. The establishment of a civil commission that has authority to review all records. That all files from top-secret to classified must be made available, upon request, for review. That this commission will have authority to see if files are really top-secret or just hiding illegal acts. That the commission be made up of unemployed workers, teachers, or persons with tenure and retired persons. Any person who cannot easily be threatened with job loss. This is one tool of evil and corrupt people. The politicians and lobby groups cannot get away with destroying our country and world anymore. The commission will have the authority to hold the hearings and order investigations.

American Nazi's

After 9-11, our intelligence community has increased by two thousand agencies and 840,000 persons. My struggles in researching my books has caused great interest in this community. Many questionable actions have taken place against me and mine.

In December 2009, 400+ documents were stolen from me, in a house in Staten Island, NY. False statements entered on my credit report. One being a child-support claim, entered before I requested a loan for a land deal. I called the childsupport center and they stated that they did not enter anything against my credit report. The credit agency refused to remove the claim.

Two entrees were made on my credit by two credit agencies, at the same time. Each demanding payment for an outstanding Verizon account. If we check my credit history, you will find I pay my bills (Thank God). Taxes on my investment property were raised from three hundred dollars a quarter to three thousand. This taking place during the release of the first book. A twenty thousand dollar credit card was reduced to one thousand dollars; during my land deal.

I would get four flat tires on my family cars within a one-year period (I think). My sideview mirror on my work car would be blasted of in the middle of the night. My daughter's car would get hit twice in the rear left side of the car, one time blasting off the bumper. This car with my documents in it would be stolen. This car had no rear bumper and almost ten years old.

Ever time I would get a letter from the NSA, a Verizon truck would be parked in front of my house. My computers would get infected twelve to fourteen times in the past four years. I remember getting one to two viruses in a four-year period. My computer would appear to be controlled from another location. Sometimes I would type and nothing happens, I would see on the bottom of the screen (copying).

Many actions to be just life's chances and bad luck. So if secret American Nazi's do exist, how do we stop them? The civil commission and investigations maybe. Many more actions have taken place, but we will watch and pray for justice. I hope many other people will stand up for our freedoms and civil liberties. I have presented some of my proof on the next pages.

Noam Chomsky

I have been in communication with
Noam Chomsky. Trying to get him
To help in this cause. His computer
Crashed the same week mines did.
My sister was helping me edit the
First book and her computer was
Infected with a virus also.

05~~0000021ab-3
ws Eudora
Date: Wed, 03 Mar 2010 16:02:24 -0500
To: chomsky@mit.edu
From: Noam Chomsky <chomsky@MIT.EI
Subject: Public talks in March, 2010
C~.

X-neerallow" true
Y-:: AAAAAwAAAAoTIxCLEyPwVA==

Our apologies for the form letter.

We used to send out automatic monthly listings, but no longer do so, for a variety of reasons. It's best if you e-mail us at this address at the beginning of each month for information on public talks scheduled for that month (as most of you have done!), as our schedule changes quickly.

We apologize if this message does not address your specific area, but we send out one e-mail to all inquiries, and it includes all public talks for that month, regardless of location.

Below are the public talks for the month of March. Apologies to those of you who did not receive info on the March 2 BU event. Our computer had some crashing issues, unfortunately, and so messages were not sent. As a bonus, I'll include here info on a Howard Zinn event which is yet to be confirmed...keep in touch.

I will also add an MIT event for Friday, April 2, since it is so close to the beginning of the month.

Asst. to Noam Chomsky

****

Saturday, March 6

Memorial Church, Cambridge, MA (Harvard Yard in Harvard Square)
12:00-1:30 pm   Event for Harvard Extension International Relations Club
                    With Amy Goodman

Topic: Current policies of the Obama administration and US foreign policy

Location: Memorial Church

Harvard University
Cambridge, MA

Contact:
Jefn r

****

Monday, March 8
Princeton University
Princeton, NJ

7:30 pm   25-30 min lecture, followed by 30-40 -min Q&A
           Noam chooses area of Said's work for lecture topic
           Title: "'I am Kinda': Reflections on the Culture of Imperialism."

Contact:

****

Thursday, March 11
MIT Model UN

6:00 pm   30-min introductory lecture
           Topic: Views on the efficacy of the United Nations,
           possible improvements, and the fairness of veto power
           in the Security Council.

6:30 pm   45-min Q&A

gx=1&.rand=5lc02pisgjoln   3/7/2010

Customer Copy

Your Work Order Number is 1257659974 Print Return
to Menu
7/2/2009 6:19:17 PM
NY Electronics Repair License # 1229379   NY Electronics Store License
1241326

----In Store Service Request----

Store Number	1257	Associate Name	Jimmy
Store Address	409 Gateway Drive Gateway Center Brooklyn ,NY 11239	Store Phone #	(718)348-9477

Customer Information

Today's Date	7/2/2009		
Customer First Name	Gil	Customer Last Name	Roman
Address			
City, State	Queens,NY	Zip Code	11416
Customer Home Phone #		Customer Business Phone #	
Email			
Items left by Customer	N/A , and the antivirus Kaspersky.		
Products purchased by Customer	N/A		

Equipment Information

Service For		Case Number	
Product	Laptop	If Other (product)	
Brand	MISCELLANEOUS SALES	Model	IBM thinkpad R50e
Boot Password		Did Machine Boot Properly Y/N	Y
Permission to Format/Recover?	Y	Serial Number	not visible
Staples Perform Backup?	N		
Description of Services	The product key of this computer is hv7jv-ybvBh-ckxjv-q6jdd-3mm4m the actual serial number was not visible therefore i took the product key number. Customer wants a virus spyware removal along with restoring the system to factory settings and the installation of the antivirus that he brought in to the store which is called Kaspersky. If needed the user name is sroma865 and the password is SamieGirl4		
Physical Damage	No Physical Damage.		

Service Description	Sku #	Unit Cost	QTY	Extended Cost
Virus/Spyware/Adware Removal	616677	$89.99	1	$89.99
System Restore	567125	$69.99	1	$69.99
Software Install (excluding OS)	860650	$29.99	1	$29.99
Subtotal:				$189.97

Grandtotal:	$189.97

Drop Off Date _____ Pickup Date _____

Staples recommends that you back up all of the information on your computer's hard drive prior to any installation or
service. You understand and agree that prior to any service performed on your computer, it is your responsibility to make

http://easytech.staples.com/easytech/ConfirmWithTerms.asp?rpo=1257659974&pswrid=...   07/02/2009

verizon

Bill Date May 28, 2009
Account 0093325085647
Invoice 61450860

Page 1 of 2

GILBERT ROMAN

Account Summary

Previous Charges	$29.99
Payment Received Thank You	-$29.99
Past Due Charges	**$.00**
New Charges	
Verizon Broadband Services	$29.99
Total New Charges	**$29.99**
Total Due	**$29.99**

DSL # 718-296-3416

June 17 No internet
service over
2 weeks
4 weeks
~~total cost~~

My computers
and under constant
attack ←
12-14 times infected or crashed

Bank of America

Platinum
Visa Business Card
Company Statement

Credit Limit	$20,000	Billing Date	05-05-09
Cash Limit	$10,000	Days in Billing Cycle	30
Cash Advance Balance	$0.00	Payment Due Date	05-30-09
Available Credit	$19,525	Minimum Payment Due	$12.94
		New Balance	**$474.97**

MAXIMINOS GENERAL CONTRACTING

Company Account Number:

Page 1 of 2

COMPANY SUMMARY

MAXIMINOS GENERAL CO 4339 9300 1985 6097	Previous Balance	- Payments	- Credits	+ Purchases/Other Debits/Fees	+ Cash Advances	+ Finance Charges	= New Balance
Company Total	$1,066.70	$600.00	$0.00	$0.00	$0.00	$8.27	$474.97

CARDHOLDER NEW ACTIVITY SUMMARY

	Credits	Purchases and Other Debits	Cash Advances	Total Activity
GILBERT ROMAN				
Credit Limit $20,000	$600.00	$0.00	$0.00	$600.00CR

Customer Service 800.673.1044, 24 hours	Finance Charges	Total Annual Percentage Rate			12.24%	**Company Account Summary**		
		Average Daily Balance	Daily Periodic Rate	Annual Percentage Rate	Periodic Finance Charge	Previous Balance		$1,066.70
Outside the U.S. 509.353.6656, 24 hours	PURCHASES	$0.00	0.03354%	12.24%	$0.00	Payments	-	$600.00
	CASH	$0.00	0.05477%	19.99%	$0.00	Credits	-	$0.00
For Lost or Stolen Card: 800.673.1044, 24 hours	PROMO 1	$821.36	0.03354%	12.24%	$8.27	Purchases/Other Debits/Other Fees	+	$0.00
						Cash Advances	+	$0.00
						Overlimit Fees	+	$0.00
						Late Payment Fees	+	$0.00
						Finance Charge	+	$8.27
						New Balance	=	$474.97

Send Billing Inquiries to:
BANK OF AMERICA
PO BOX 15184
WILMINGTON DE 19850-5184

Please see the reverse side for information about your account.

Please return coupon with your payment.

Business Card Payment Coupon

☐ Check box and indicate address change on reverse.

Company Account No.	
Payment Due Date	05-30-09
Minimum Payment Due	$12.94
New Balance	$474.97

Please Enter Amount Enclosed $ _____

Make check or money order payable to:
BUSINESS CARD
Mail payment to address below.

MAXIMINOS GENERAL CONTRACTING   **P0031627

OZONE PARK NY 11416-1107

BUSINESS CARD
PO BOX 15710
WILMINGTON DE 19886-5710

4339930019856097000129400474 97

1: 5499900 1 1: 000 300 198 56097 1

84

Platinum
Visa Business Card
Company Statement

Credit Limit	$1,000	Billing Date	06-05-09
Cash Limit	$200	Days in Billing Cycle	31
Cash Advance Balance	$0.00	Payment Due Date	06-30-09
Available Credit	$1,000	Minimum Payment Due	$0.00
		New Balance	**$0.08CR**

MAXIMINOS GENERAL CONTRACTING

Company Account Number:

COMPANY SUMMARY

MAXIMINOS GENERAL CO 4339 9300 1985 6097	Previous Balance	- Payments	- Credits +	Purchases/Other Debits/Fees +	Cash Advances +	Finance Charges =	New Balance
Company Total	$474.97	$479.39	$0.00	$0.00	$0.00	$4.34	$0.08CR

CARDHOLDER NEW ACTIVITY SUMMARY

	Credits	Purchases and Other Debits	Cash Advances	Total Activity
GILBERT ROMAN				
Credit Limit $1,000	$479.39	$0.00	$0.00	$479.39CR

Customer Service
800.673.1044, 24 hours

Outside the U.S.
509.353.6656, 24 hours

For Lost or Stolen Card:
800.673.1044, 24 hours

Send Billing Inquiries to:
BANK OF AMERICA
PO BOX 15184
WILMINGTON DE 19850-5184

Finance Charges	Total Annual Percentage Rate		12.24%
	Average Daily Balance	Daily Periodic Rate	Annual Percentage Rate	Periodic Finance Charge
PURCHASES	$0.00	0.03354%	12.24%	$0.00
CASH	$0.00	0.05477%	19.99%	$0.00
PROMO 1	$416.66	0.03354%	12.24%	$4.34

Company Account Summary		
Previous Balance		$474.97
Payments	-	$479.39
Credits		$0.00
Purchases/Other Debits/Other Fees	+	$0.00
Cash Advances	+	$0.00
Overlimit Fees	+	$0.00
Late Payment Fees	+	$0.00
Finance Charge	+	$4.34
New Balance	=	$0.08CR

Please see the reverse side for information about your account.

Please return coupon with your payment.

Business Card Payment Coupon

☐ Check box and indicate address change on reverse.

Company Account No.

Payment Due Date	06-30-09
Minimum Payment Due	$0.00
New Balance	$0.08 CR

Please Enter Amount Enclosed $ _____

Make check or money order payable to:
BUSINESS CARD
Mail payment to address below.

MAXIMINOS GENERAL CONTRACTING
**P0047716
OZONE PARK NY 11416-1107

BUSINESS CARD
PO BOX 15710
WILMINGTON DE 19886-5710

4339930019856097000000000000008

:549900 1 1:000300 198 56097"

January 27, 2010

Constant computer attacks caused
Me not be able to enter this case
To the Supreme Court of the US.
My computer would not respond
To my commands and conform
To booklet regulations.

Gilbert Roman

Ozone Park, NY 11416

RE: Roman v. NSA

Dear Mr. Roman:

The above-entitled petition for a writ of certiorari was postmarked January 21, 2010 and received January 26, 2010.

The papers are returned for the following reason(s):

If you intend to pay the $300.00 docket fee, the petition must be in booklet format and on paper that measures 6 1/8 by 9 1/4 inches and comply with the filing requirements of Rule 33.1.

However, if you wish to proceed in forma pauperis, you must also submit a motion for leave to proceed in forma pauperis and a notorized affidavit or declaration of indigency. You may use the enclosed form. Rule 39. You need only to provide an original and 10 copies of the petition and motion for leave to proceed in forma pauperis. Rule 12.2.

Please correct and resubmit as soon as possible. Unless the petition is received by this Office in corrected form within 60 days of the date of this letter, the petition will not be filed. Rule 14.5.

A copy of the corrected petition must be served on opposing counsel.

Your $300 check #339 is herewith returned.

Sincerely,
William K. Suter, Clerk
By:

S. Elliott
(202) 479-3025

Enclosures

3023

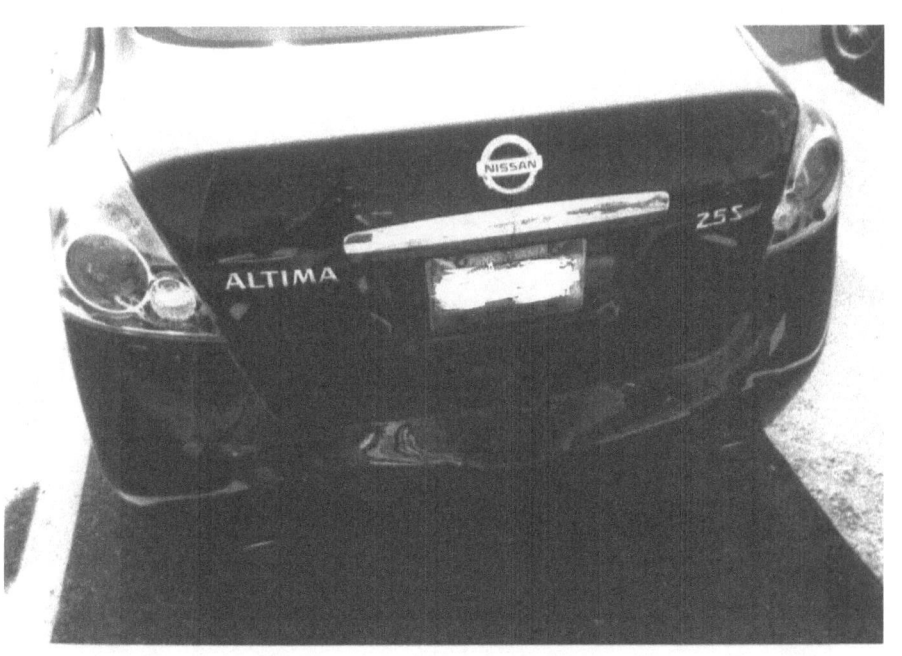

NSPECTION STATION   MOTOR VEHICLE REP.

& Collision Inc.

Copiague, New York 11726

NAME __Gilbert ROMAN__

ADDRESS

__Ozone Plc. NY 11416__

YEAR & MAKE OF CAR __2008 Nissan__

LICENSE PLATE NO. ____

PHONE ____   DATE __9-4-09__

	AMOUNT	
	PARTS	LABOR
Repair L/s qt panel *450 —		
+ Refinish		
Replace rear bumper + Install + refinish *1,350		
Parts + Labor		
450.00		

88

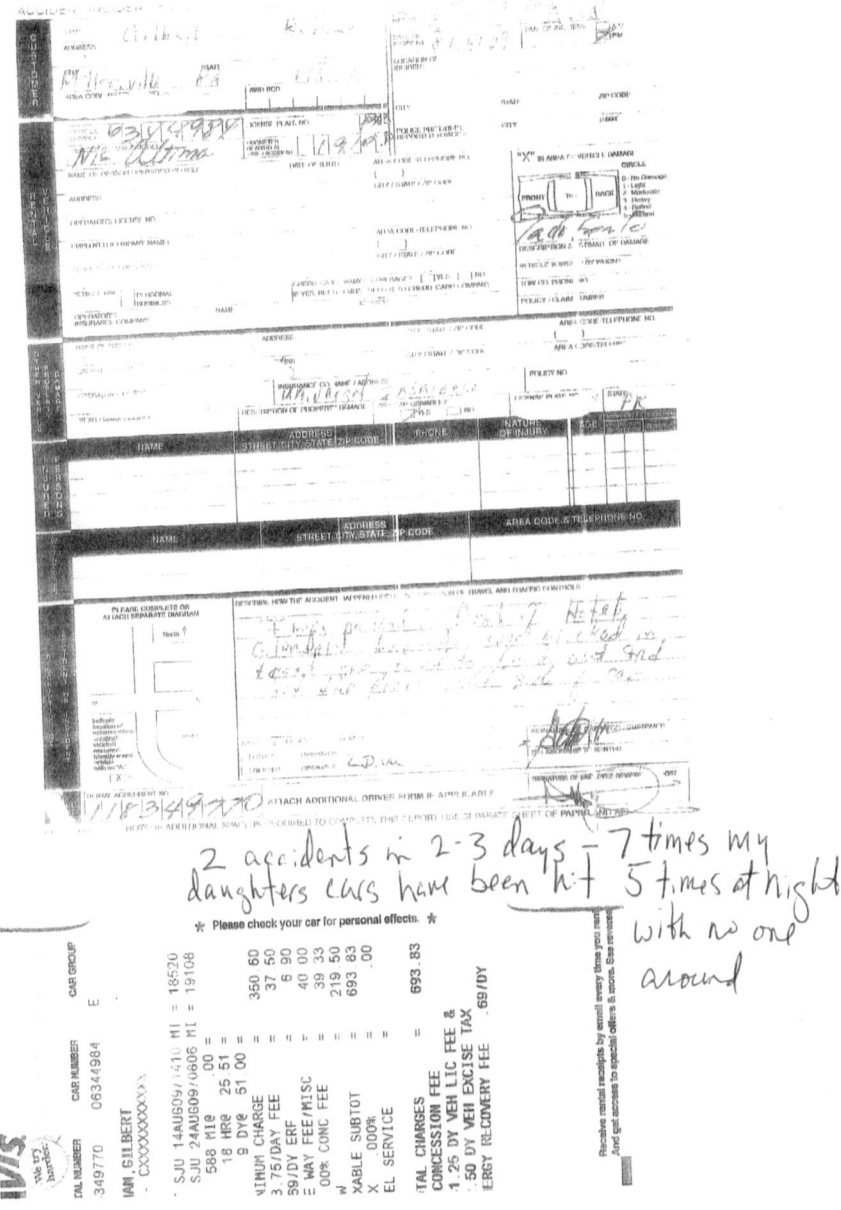

2 accidents in 2-3 days — 7 times my
daughters cars have been hit 5 times at night
with no one
around

```
          TROPICAL FOOTPRINTS #3
          180 EL CONQUISTADOR HOT
             FAJARDO, PR 00738


     TIME  11:13 AM     DATE  08/17/09
     TERMN 0001          MERN 679/4430027
     TRAN TYPE SAL+CASH
     #xxxxxxxxxxxxL
     CARD TYPE DEBIT.

     TRAN # 142


     ACCT TYPE CHECKING RECEIPT # 000462

     TOTAL              ********
```


```
          MERCHANT COPY
```

My debt cards and credit cards are repeatedly not working? This is just one (one vacation in Puerto Rico).

Two times in 2010, I tried to pay an ad fee for *Poets and Writers* magazine, and I had over three thousand dollars in the bank, and the debt card was declined.

I have here three complaints with the post office.

Oh, my mail is being opened, and I am receiving it from slightly opened. Two grossly opened, one letter to Art Bell from Coast to Coast Am Radio looked like several people opened it.

03/28/2008   Trans**Union**.

POHEAP00205111-I028739
GILBERT ROMAN

IᵢᵤIIₗᵤIIIIₗₗᵤIIₗᵤᵤIIₗᵤIIₗIₗᵤₗₗIₗIIIₗᵤIᵤIₗIIₗᵤIₗI

Our investigation of the dispute you recently submitted is now complete. The results are listed below.

If our investigation has not resolved your dispute, you may add a 100-word statement to your report. If you provide a consumer statement that contains medical information related to service providers or medical procedures, then you expressly consent to TransUnion including this information in every credit report we issue about you.

If there has been a change to your credit history resulting from our investigation, or if you add a consumer statement, you may request that TransUnion send an updated report to those who received your report within the last two years for employment purposes, or within the last one year for any other purpose.

If interested, you may also request a description of how the investigation was conducted along with the business name, address and telephone number of any company we may have contacted for information.

Thank you for helping ensure the accuracy of your credit information.

Investigation Results

ITEM	DESCRIPTION	RESULTS
PERSONAL INFORMATION		NEW INFORMATION BELOW
DEBT RECOVERY SOLUTIONS	# 7187638152103	VERIFIED, NO CHANGE
DIVISION OF CHILD SUP EN	# BG218220D2	NEW INFORMATION BELOW

P 0HEAP-002 05111-I028739 01/08

92

Personal Information

Name: GILBERT ROMAN

SSN:
Date of Birth:
Telephone:
Your SSN is partially masked for your protection.

You have been on our files since 01/2003

CURRENT ADDRESS

Address:

Date Reported: 02/2008

PREVIOUS ADDRESS

Address:
                                        , NJ 07062

Date Reported: 10/2007

Address:
                    STROSSER, NY 11791

EMPLOYMENT DATA REPORTED

Employer Name: AT&T
Date Reported: 04/2007

Position: INSPECTOR
Hired:

Employer Name:
Date Reported: 07/2005

Position:
Hired:

Employer Name:
Date Reported: 01/2004

Position: PLUMBER
Hired:

Special Notes: If any item on your credit report begins with 'MED1', it includes medical information and the data following 'MED1' is not displayed to anyone but you except where permitted by law.

Account Information

The key to the right helps explain the payment history information contained in some of the accounts below. Not all accounts will contain payment history information, but some creditors report how you make payments each month in relation to your agreement with them.

N/A	X	OK	30	60	90	120
Not Applicable	Unknown	Current	30 days late	60 days late	90 days late	120 days late

Adverse Accounts

The following accounts contain information which some creditors may consider to be adverse. Adverse account information may generally be reported for 7 years from the date of the first delinquency, depending on your state of residence. The adverse information in these accounts has been printed in >brackets< or is shaded for your convenience, to help you understand your report. They are not bracketed or shaded this way for creditors. (Note: The account # may be scrambled by the creditor for your protection).

DEBT RECOVERY SOLUTIONS #7187638152103

900 MERCHANTS CONC
SUITE 106
WESTBURY, NY 11590
(516) 228-7110

Balance: $87
Date Verified: 02/2008
Original Amount: $87
Original Creditor: VERIZON
Past Due: >$87<

Pay Status: >COLLECTION ACCOUNT<
Account Type: OPEN ACCOUNT
Responsibility: INDIVIDUAL ACCOUNT
Date Closed: 11/2007

Loan Type: FACTORING COMPANY ACCOUNT
Remarks: ACCT INFO DISPUTED BY CONSUMR
Date placed for collection: 08/2007

P 0HEAP-002 05111-I028741 03/08

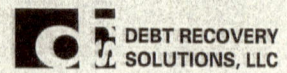 **DEBT RECOVERY SOLUTIONS, LLC**

September 13, 2007

Gilbert Roman

900 MERCHANTS CONCOURSE, SUITE 106
WESTBURY, NEW YORK 11590-5114
(516) 228-7110 / 1-800-851-9758

Balance Due   : $87.79
Original Creditor : VERIZON
Account #   : 1247187638152103
Original Phone #

Your account was acquired by Debt Recovery Solutions through a purchase of various delinquent receivables from VERIZON. All communication, correspondence and payments on this account should be directed to this office. This initial letter is the beginning of our collection process designed to impress upon you the importance of resolving this long - standing debt. This delinquency is an issue that you should review very carefully since planning for the future can often be affected by your past activities.

Unless you notify this office within 30 days after receiving this notice that you dispute the validity of this debt or any provision thereof, this office will assume this debt is valid. If you notify this office of any dispute in writing within 30 days after receiving this notice, this office will obtain a verification of the debt or obtain a copy of a judgment and mail you a copy of such verification or judgment. If you request this office in writing within 30 days of receiving this notice, this office will provide you with the name and address of the original creditor, if different from the current creditor.

As required by law, you are hereby notified that a negative credit report reflecting on your credit record has been submitted to a credit reporting agency because you have failed to fulfill the terms of your credit obligations.

John Donnell, Recovery Specialist
1-800-851-9758
(Mon-Thur 9:00 A.M. -10:00 P.M., Fri 9:00 A.M.-3:30 P.M. & Sat 9:00 A.M.-1:30 P.M.EST)

New York City Department of Consumer Affairs license number 1114291.

 ACA
INTERNATIONAL
The Association of Credit
and Collection Professionals

This is an attempt to collect a debt and any information obtained will be used for that purpose. This communication is from a debt collector.

NOTICE: See Reverse Side for Important Information.
------------------------------- Detach and Return with Payment -------------------------------
☐ Please check here if there is a new phone number or address and provide the information on the reverse side.

PO BOX 1259
Oaks, Pa 19456

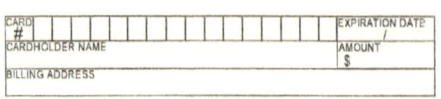

Statement Of Account

Balance Due   : $87.79
Original Creditor : VERIZON
Account #   : 1247187638152103
Original Phone #   : (718) 763-8152

◀ If you wish to pay by Visa or MC complete this information

CARD #											EXPIRATION DATE /
CARDHOLDER NAME										AMOUNT $	
BILLING ADDRESS											

1085 - 418

Gilbert Roman

DEBT RECOVERY SOLUTIONS, LLC
P.O. BOX 9001
WESTBURY, NEW YORK 11590-9001

DL1   1085 - 418   DR101-4

94

Afni, Inc.
PO Box 3427
Bloomington, IL 61702-3427
(888)257-1585
www.afnicollections.com

Megan

COLLECTION NOTICE

This account has been acquired by our agency for collection. We believe it is in your best interest to resolve this account. We may report information about your account to credit bureaus.

If you have any questions, please contact our office toll free at (888)257-1585 Monday through Friday 7am-9pm CST. For proper credit on your account, please write this number 027326446-02 on your payment.

Unless you notify this office within 30 days after receiving this notice that you dispute the validity of the debt or any portion thereof, this office will assume this debt is valid. If you notify this office in writing within 30 days from receiving this notice that you dispute the validity of the debt or any portion there of, this office will: obtain verification of the debt or obtain a copy of a judgement and mail you a copy of such judgement or verification. If you request this office in writing within 30 days after receiving this notice, this office will provide you with the name and address of the original creditor, if different from the current creditor. This is an attempt to collect a debt. Any information obtained will be used for the purpose. You have the right to inspect your credit. This letter is from a debt collector.

Please see reverse side of this notice for our Privacy Statement and credit card payment options.

To manage your account online, visit us at : **www.afnicollections.com**

New York City Department of Consumer Affairs License Number #1072175

This account is for a remaining balance from the original creditor for services associated with the previous telephone number listed below.

Please retain this information for your records

Afni, Inc. Account #	Balance Due	Original Creditor	Disconnected Phone #	Date
027326446-02	$288.14	Verizon New York Inc.	(516)564-4489	3/18/2008

Detach along perforation and return bottom portion along with payment in the enclosed envelope. Credit card payment options are on the back of notice.

For proper credit, please include your Afni account # listed below on your check

AFNFF-0318D523129-QWSP-2 3125

Department 555
PO BOX 4115
CONCORD CA 94524

ADDRESS SERVICE REQUESTED

#BWNFTZF #AFN8082255308034#

GILBERT ROMAN

_____ AHK NY 11416-1107

Afni, Inc. Account #:   027326446-02
Original Creditor: Verizon New York Inc.
Disconnected Phone #: (516)564-4489
Balance Due: $288.14
Date: 3/18/2008
Toll Free: (888)257-1585

9 ' 02027326446   991700   28814

PO Box 3427
Bloomington, IL 61702-3427

November 18, 2008

Dear GILBERT ROMAN:

If you are a victim of fraud, the first step in protecting your credit information is to add a fraud alert to each of the credit files maintained by the three national credit reporting agencies. Adding a fraud alert may aid in the prevention of further fraudulent activity.

We were successful in adding an alert to your Equifax credit file. We will forward your information to Experian and TransUnion and they will also add an alert to your credit file in their databases, eliminating the need for you to contact each credit reporting agency directly.

The alert will remain on your Equifax credit file for 90 days. This alert may help protect the credit file from being used fraudulently.

You may request a free copy of your credit file. This can be done by visiting www.equifax.com/fcra, by calling **1-800-685-1111** and using the automated ordering system *OR* you may submit your request in writing to:

Equifax Information Services LLC
P.O. Box 105069
Atlanta, GA 30348-5069

Please order your credit file within 90 days of the date of this letter.

If you do not receive notification from Experian and from TransUnion that they have added a fraud alert for you on their credit files, please contact them directly using the following contact information:

TransUnion, PO Box 6790, Fullerton, CA 92634 : (800) 680-7289

Experian, PO Box 9530, Allen, TX 75013 : (888) 397-3742

Thank you for the opportunity to assist you.

Equifax Information Services LLC

The FBI has named identity theft as the fastest growing crime in America.

Protect yourself with Equifax Credit Watch™ , a service that monitors your credit file every business day and notifies you within 24 hours of any activity. To order, go to: *www.creditwatch.equifax.com*

Someone else opened two bank
Accounts in my name, on the
Internet. I ordered these accounts
Closed.

11/19/2008

GILBERT ROMAN

OZONE PARK , NY 11416-1107

Re. Account(s) XXXXXXX7959, XXXXXXX7974

Dear GILBERT ROMAN:

We regret to inform you that Bank of America has elected to close your account(s) in
accordance with the provisions of our Deposit Agreement and disclosures provided to
you at the time your account(s) was opened. Under these terms and conditions, either
the bank or the customer may close the account(s) at any time.

Please make other banking arrangements for the handling of any automatic and/or
electronic transactions, and do not write any checks.

When the account(s) is closed, any checks presented for payment will be returned
'Account Closed' and if you have an ATM/Check Card it will no longer access the
account. A Cashier's Check for any collected balance will be mailed to you after all
previously deposited items have been verified.

If your account(s) is overdrawn or becomes overdrawn, a deposit of cash must be made
to bring the account(s) to a zero balance. Additionally, we may report the account(s)
to Chex Systems, Inc., an account verification service. This may adversely impact
your ability to open an account at another financial institution for up to five years.

If you have any questions about this matter, please contact Risk Identification
Support Center Customer Service at 1.877.240.6886 Option 2 between 8:00 AM and 06:00
PM Eastern Time Monday through Friday, and 9:00AM and 5:00PM Eastern on Saturday.

Thank You,
Customer Service

Case Ref#:   25451904

CARDMEMBER SERVICE
P.O. BOX 15299
WILMINGTON, DE 19850-5299
FAX: (888) 643-9624

CHASE ◒

January 27, 2009

Ilı,ıllıılılalıllıılllıılllılılıllllılllılılllıllll

13048 RCS 001 011 02709 - NNNNNNNNNNNN
Gilbert Roman

Ozone Park NY 11416-1107

RE: Your account
ending in

Dear Gilbert Roman,

We are investigating your dispute about a 229.89 charge from
CATERING that originally appeared on your credit card account on January
07, 2009.

So we can look into this matter, we have asked the merchant's bank to send
us a copy of the sales slip. We expect to receive the sales slip within
35 days, which is the time we are required to give the merchant. After we
complete our investigation, we will write to you about how this matter was
resolved.

If you have any questions, please call us at the toll-free number shown on
the back of your credit card. For your convenience, we are available 24
hours a day to assist you.

Sincerely,

Financial Service Advisor

DSP207

AFFIRMATION OF UNAUTHORIZED USE

, GILBERT ROMAN , state and declare that I am the Cardmember of the Credit Card listed below.

Any transaction(s) occurring on or after January 6, 2009 is/are also unauthorized.

ther I, nor any authorized user of the account number indicated above, including any authorized third parties, have
...ed this credit card for the charges that are indicated by me below as unauthorized. Furthermore, neither I, nor any
authorized user of the account received any benefit or value from these charges.

Signed: _____   3-10-09
      Cardholder Signature   Date

Signed: _____   _____
      Cardholder Signature   Date

To expedite processing, you may fax this Affirmation to 866-827-7034.

Write any additional unauthorized transaction(s) on the blank line(s). If more room is needed, include the additional
unauthorized transactions on a separate sheet of paper and return with the Affirmation.

UNAUTHORIZED TRANSACTIONS
Circle "Yes" if transactions are authorized
Circle "No" if transactions are not authorized

Authorized	Date	Amount	Merchant/Bank	Address	Transaction Acct#
YES NO	01/06/2009	229.89	CATERING	5407528246 VA 224060000 US 0000000000000000	
YES NO					
YES NO					
YES NO					
YES NO					
YES NO					
YES NO					
YES NO					
YES NO					
YES NO					
YES NO					

Date: February 27, 2009   **RISBDM**

0010000065983827001

RESERVATION REWARDS

Customer Service
P.O. Box 855
Shelton, CT 06484

Someone else opened this account in my name.

Gilbert Roman

Ozone Park, NY 11416-1107
||..||..||.|.|..|||.||..||.||..||||||..|..|.|.|.|.|..|||

July 15, 2009

Dear Gilbert Roman,

As part of a review of our Reservation Rewards membership files, we ask that you please confirm the accuracy of the information we currently have on record:

Membership in Name of:	ꞏ...
Membership Number:	26464136000
Billing Address:	
	Ozone Park, NY 11416-1107
Member Since:	07/13/2009
Joined Through:	Continental Airlines
Coupon/Promotion Offered:	$20.00 Cash Back Incentive on next Continental Airlines Reservation

If the membership information above is accurate, you do not need to respond to this notice.

Reservation Rewards provides its members with access to hundreds of dollars of savings per year with discounts on dining, shopping, entertainment and more. The $12 per month membership fee for this account is being billed to your Visa credit or debit card with the last four digits ꞏ ꞏ

If the membership information above is NOT accurate or if you have questions about this account, please call our Customer Service Team at 1-800-732-7031, Mon – Fri 8 AM – 11 PM (ET), Sat 9 AM – 6 PM (ET) or Sun 9 AM – 5 PM (ET).

Thank you for your assistance.

Sincerely,

The Reservation Rewards Customer Service Team

BAT1-RR-07152009-0000053

100

FRANKLIN TOWNSHIP   SOMERSET COUNTY

BLOCK NUMBER	LOT NUMBER	QUALIFICATION
552	11	

Property Line: LEONA ST
Bldg Dims:
Additional Lots: L12
Land Dimens: 50x110
Bank: Mortgage #   Tax Acct #

ASSESSED VALUATION INFORMATION
▼ LAND ▼   ▼ IMPROVEMENTS ▼   ▼ TOTAL ▼

EXEMPTIONS ►   NET TAXABLE VALUE ►

08846

DESCRIPTION	AMOUNT OF TAX
MUNICIPAL TAX:	3.18
NON-MUNICIPAL TAX:	16.33
(SUBTOTAL):	19.51
NET TAX:	19.51

" PAST DUE AMOUNT "

2008 2ND QUARTER DUE MAY 1, 2009	2009 1ST QUARTER DUE FEB 1, 2009
9.75	9.76

INFORMATION FOR TAXPAYERS

MAKE CHECK
PAYABLE TO: TOWNSHIP OF FRANKLIN

MAIL TO: COLLECTOR OF REVENUE
TOWNSHIP OF FRANKLIN
PO BOX 5059
SOMERSET, NJ 08875

SEE REVERSE SIDE FOR ADDITIONAL INFORMATION

REFER INQUIRIES REGARDING
COUNTY TAXES TO RICHARD WILLIAMS
(908) 231-7000.
SCHOOL TAXES TO EDWARD SETO
(732) 873-2400.
MUNICIPAL TAXES TO KENNETH DALY
(732) 873-2500.

TAX INQUIRIES (732) 873-2500 x337

2009 1st & 2nd Quarter Tax Bill

DISTRIBUTION OF TAXES

Municipal Taxes	$3.18
Non Municipal Taxes	$16.33

STATE AID USED TO OFFSET LOCAL PROPERTY TAXES: The budgets of the government agencies funded by this tax bill include State aid used to reduce property taxes. Based on the assessed value, the amount of this State aid used to offset property taxes on this parcel equals:

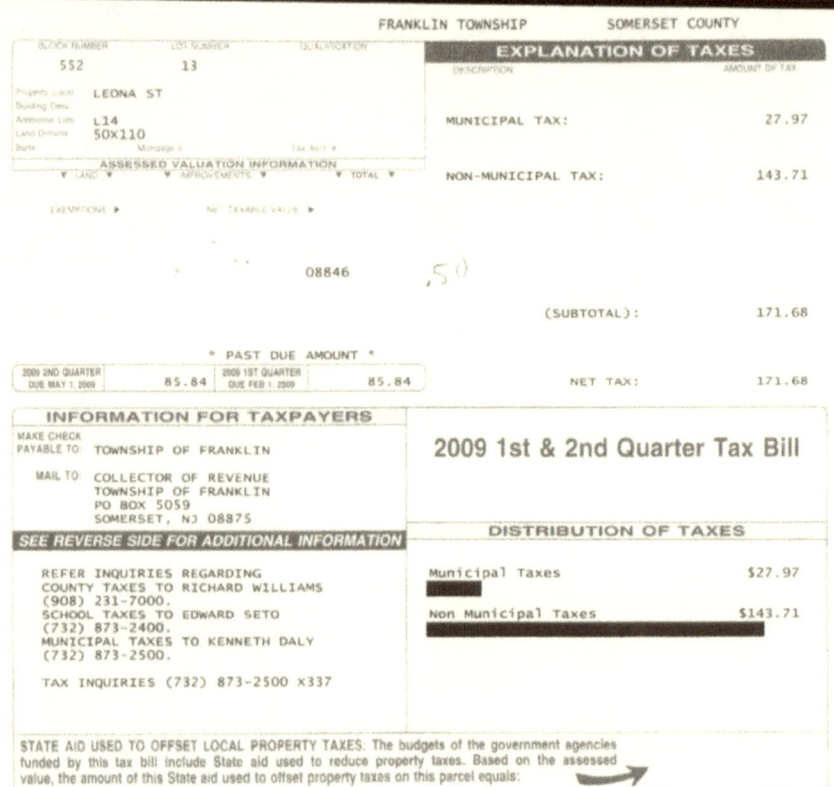

FRANKLIN TOWNSHIP   SOMERSET COUNTY

BLOCK NUMBER	LOT NUMBER	QUALIFICATION
552	13	

EXPLANATION OF TAXES

DESCRIPTION	AMOUNT OF TAX

Property Loc: LEONA ST
Building Desc:
Additional Lots: L14
Land Dimens: 50x110

MUNICIPAL TAX: 27.97

ASSESSED VALUATION INFORMATION
▼ LAND ▼   ▼ IMPROVEMENTS ▼   ▼ TOTAL ▼

NON-MUNICIPAL TAX: 143.71

EXEMPTIONS ▶   NET TAXABLE VALUE ▶

08846   ,50

(SUBTOTAL):   171.68

* PAST DUE AMOUNT *

2009 2ND QUARTER DUE MAY 1, 2009	85.84	2009 1ST QUARTER DUE FEB 1, 2009	85.84

NET TAX: 171.68

INFORMATION FOR TAXPAYERS

MAKE CHECK
PAYABLE TO: TOWNSHIP OF FRANKLIN

MAIL TO: COLLECTOR OF REVENUE
TOWNSHIP OF FRANKLIN
PO BOX 5059
SOMERSET, NJ 08875

SEE REVERSE SIDE FOR ADDITIONAL INFORMATION

REFER INQUIRIES REGARDING
COUNTY TAXES TO RICHARD WILLIAMS
(908) 231-7000.
SCHOOL TAXES TO EDWARD SETO
(732) 873-2400.
MUNICIPAL TAXES TO KENNETH DALY
(732) 873-2500.

TAX INQUIRIES (732) 873-2500 x337

2009 1st & 2nd Quarter Tax Bill

DISTRIBUTION OF TAXES

Municipal Taxes $27.97

Non Municipal Taxes $143.71

STATE AID USED TO OFFSET LOCAL PROPERTY TAXES: The budgets of the government agencies
funded by this tax bill include State aid used to reduce property taxes. Based on the assessed
value, the amount of this State aid used to offset property taxes on this parcel equals:

FRANKLIN TOWNSHIP   SOMERSET COUNTY

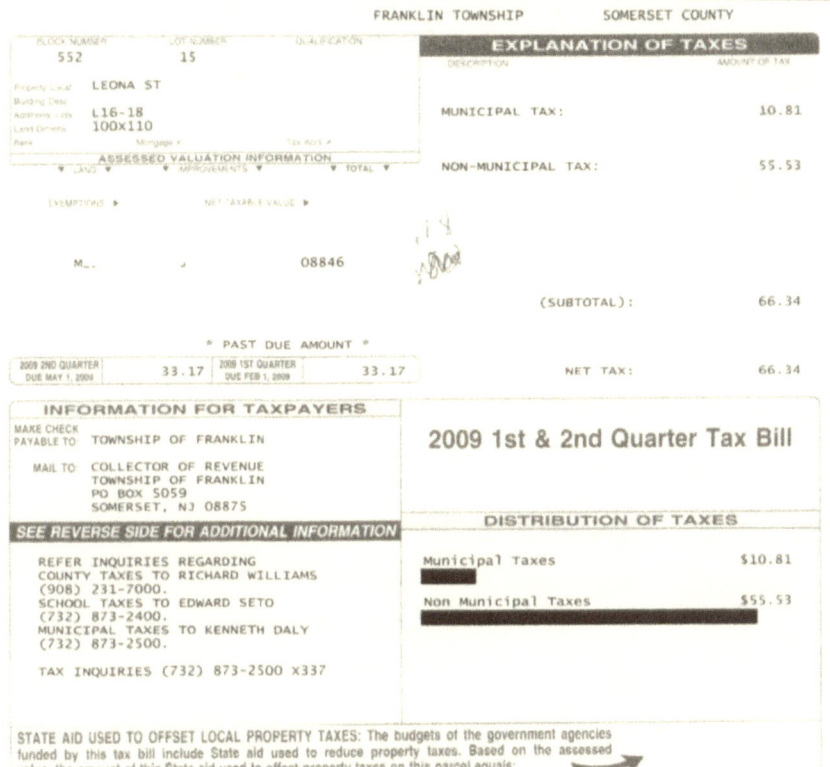

BLOCK NUMBER	LOT NUMBER	QUALIFICATION
552	15	

Property Local   LEONA ST
Building Desc
Additional Lots   L16-18
Land Dimens   100x110
Bank   Mortgage #   Tax Acct #

ASSESSED VALUATION INFORMATION
LAND ▼   IMPROVEMENTS ▼   TOTAL ▼

EXEMPTIONS ▶   NET TAXABLE VALUE ▶

M_.   .   08846

* PAST DUE AMOUNT *

2009 2ND QUARTER DUE MAY 1, 2009	33.17	2009 1ST QUARTER DUE FEB 1, 2009	33.17

EXPLANATION OF TAXES
DESCRIPTION   AMOUNT OF TAX

MUNICIPAL TAX: 10.81

NON-MUNICIPAL TAX: 55.53

(SUBTOTAL):   66.34

NET TAX: 66.34

INFORMATION FOR TAXPAYERS

MAKE CHECK PAYABLE TO   TOWNSHIP OF FRANKLIN

MAIL TO   COLLECTOR OF REVENUE
          TOWNSHIP OF FRANKLIN
          PO BOX 5059
          SOMERSET, NJ 08875

SEE REVERSE SIDE FOR ADDITIONAL INFORMATION

REFER INQUIRIES REGARDING
COUNTY TAXES TO RICHARD WILLIAMS
(908) 231-7000.
SCHOOL TAXES TO EDWARD SETO
(732) 873-2400.
MUNICIPAL TAXES TO KENNETH DALY
(732) 873-2500.

TAX INQUIRIES (732) 873-2500 x337

2009 1st & 2nd Quarter Tax Bill

DISTRIBUTION OF TAXES

Municipal Taxes $10.81

Non Municipal Taxes $55.53

STATE AID USED TO OFFSET LOCAL PROPERTY TAXES: The budgets of the government agencies funded by this tax bill include State aid used to reduce property taxes. Based on the assessed value, the amount of this State aid used to offset property taxes on this parcel equals:

103

FRANKLIN TOWNSHIP   SOMERSET COUNTY

BLOCK NUMBER	LOT NUMBER	QUALIFICATION
552	15.01	

LOC
DESC
LOTS
DIMENSIONS 100X110 .25 AC
BANK   Mortgage #   Tax Acct #

ASSESSED VALUATION INFORMATION

▼ LAND ▼	▼ IMPROVEMENTS ▼	▼ TOTAL ▼
75000	0	75000
		75000

EXEMPTIONS ►   NET TAXABLE VALUE ►

07060

EXPLANATION OF TAXES

DESCRIPTION	RATE PER $100	AMOUNT OF TAX
MUNICIPAL TAX:	.303	227.25
LESS PREV.TAX		10.81
BAL. OF MUNICIPAL TAX		216.44
NON-MUNICIPAL TAXES:		
COUNTY TAX	.265	198.75
COUNTY OPEN SPACE TAX	.030	22.50
DISTRICT SCHOOL TAX	1.158	868.50
MUNICIPAL OPEN SPACE T	.050	37.50
FIRE DIST 1	.051	38.25
TOT NON-MUNICIPAL TAX	1.554	1165.50
LESS PREV.TAX		55.53
BAL NON-MUNICIPAL TAX		1109.97
TOTAL TAX		1326.41

2009 4TH QUARTER DUE NOV. 1, 2009	663.20	2009 3RD QUARTER DUE AUG. 1, 2009	663.21	BALANCE OF TAX DUE	1326.41

INFORMATION FOR TAXPAYERS

MAKE CHECK
PAYABLE TO: TOWNSHIP OF FRANKLIN

MAIL TO: COLLECTOR OF REVENUE
TOWNSHIP OF FRANKLIN
PO BOX 5059
SOMERSET, NJ 08875

SEE REVERSE SIDE FOR ADDITIONAL INFORMATION

REFER INQUIRIES REGARDING
COUNTY TAXES TO RICHARD WILLIAMS
(908) 231-7000.
SCHOOL TAXES TO EDWARD SETO
(732) 873-2400.
MUNICIPAL TAXES TO KENNETH DALY
(732) 873-2500.
3RD QUARTER TAX DUE BY 8/28/09 WITH
NO ADDITIONAL GRACE PERIOD.

2009 3rd & 4th Quarter Tax Bill

DISTRIBUTION OF TAXES

County Taxes	18.54%	$221.25
School Taxes	62.37%	$868.50
Municipal Taxes	16.33%	$227.25
Other Taxes	2.76%	$38.25

STATE AID USED TO OFFSET LOCAL PROPERTY TAXES: The budgets of the government agencies funded by this tax bill include State aid used to reduce property taxes. Based on the assessed value, the amount of this State aid used to offset property taxes on this parcel equals: $261.75

FRANKLIN TOWNSHIP
SOMERSET COUNTY   **2009-4**
TAX COLLECTOR'S STUB - DETACH AND RETURN WITH YOUR CHECK
2009 4TH QUARTER TAX DUE NOVEMBER 1, 2009

BLOCK NUMBER	LOT NUMBER	QUALIFICATION	BANK CODE
552	15.01		

TAX ACCOUNT NUMBER	TAX BILL NUMBER	TAX AMOUNT BILLED	DUE NOVEMBER 1, 2009
	000289	►	663.20

ADJUSTMENT
DUTIES/EST
CASH
CHECK
TOTAL

FRANKLIN TOWNSHIP
SOMERSET COUNTY   **2009-3**
TAX COLLECTOR'S STUB - DETACH AND RETURN WITH YOUR CHECK
2009 3RD QUARTER TAX DUE AUGUST 1, 2009

BLOCK NUMBER	LOT NUMBER	QUALIFICATION	BANK CODE
552	15.01		

TAX ACCOUNT NUMBER	TAX BILL NUMBER	TAX AMOUNT BILLED	DUE AUGUST 1, 2009
	000289	►	663.21

ADJUSTMENT
INTEREST
CASH
CHECK
TOTAL

FRANKLIN TOWNSHIP   SOMERSET COUNTY

BLOCK NUMBER	LOT NUMBER	QUALIFICATION
552	21	

LOC:
DESC:
LOTS:
DIMENSIONS 25x110
BANK   Mortgage #   Tax Acct. #

ASSESSED VALUATION INFORMATION

▼ LAND ▼	▼ IMPROVEMENTS ▼	▼ TOTAL ▼
18000	0	18000

EXEMPTIONS ►   NET TAXABLE VALUE ►   18000

07060

EXPLANATION OF TAXES

DESCRIPTION	RATE PER $100	AMOUNT OF TAX
MUNICIPAL TAX:	.303	54.54
LESS PREV.TAX		2.54
BAL. OF MUNICIPAL TAX		52.00
NON-MUNICIPAL TAXES:		
COUNTY TAX	.265	47.70
COUNTY OPEN SPACE TAX	.030	5.40
DISTRICT SCHOOL TAX	1.158	208.44
MUNICIPAL OPEN SPACE T	.050	9.00
FIRE DIST 1	.051	9.18
TOT NON-MUNICIPAL TAX	1.554	279.72
LESS PREV.TAX		13.06
BAL NON-MUNICIPAL TAX		266.66
TOTAL TAX		318.66

2009 4TH QUARTER DUE NOV. 1, 2009	159.33	2009 3RD QUARTER DUE AUG. 1, 2009	159.33	BALANCE OF TAX DUE	318.66

INFORMATION FOR TAXPAYERS

MAKE CHECK
PAYABLE TO: TOWNSHIP OF FRANKLIN

MAIL TO: COLLECTOR OF REVENUE
TOWNSHIP OF FRANKLIN
PO BOX 5059
SOMERSET, NJ 08875

SEE REVERSE SIDE FOR ADDITIONAL INFORMATION

REFER INQUIRIES REGARDING
COUNTY TAXES TO RICHARD WILLIAMS
(908) 231-7000.
SCHOOL TAXES TO EDWARD SETO
(732) 873-2400.
MUNICIPAL TAXES TO KENNETH DALY
(732) 873-2500.
3RD QUARTER TAX DUE BY 8/28/09 WITH
NO ADDITIONAL GRACE PERIOD.

2009 3rd & 4th Quarter Tax Bill

DISTRIBUTION OF TAXES

County Taxes	18.54%	$53.10
School Taxes	62.37%	$208.44
Municipal Taxes	16.33%	$54.54
Other Taxes	2.76%	$9.18

STATE AID USED TO OFFSET LOCAL PROPERTY TAXES: The budgets of the government agencies funded by this tax bill include State aid used to reduce property taxes. Based on the assessed value, the amount of this State aid used to offset property taxes on this parcel equals: $62.82

FRANKLIN TOWNSHIP
SOMERSET COUNTY   **2009-4**
TAX COLLECTOR'S STUB - DETACH AND RETURN WITH YOUR CHECK
2009 4TH QUARTER TAX DUE NOVEMBER 1, 2009

BLOCK NUMBER	LOT NUMBER	QUALIFICATION	BANK CODE
552	21		

TAX ACCOUNT NUMBER	TAX BILL NUMBER	TAX AMOUNT BILLED	DUE NOVEMBER 1, 2009
	000290		159.33

ADJUSTMENT

INTEREST
CASH
CHECK
TOTAL

FRANKLIN TOWNSHIP
SOMERSET COUNTY   **2009-3**
TAX COLLECTOR'S STUB - DETACH AND RETURN WITH YOUR CHECK
2009 3RD QUARTER TAX DUE AUGUST 1, 2009

BLOCK NUMBER	LOT NUMBER	QUALIFICATION	BANK CODE
552	21		

TAX ACCOUNT NUMBER	TAX BILL NUMBER	TAX AMOUNT BILLED	DUE AUGUST 1, 2009
	000290		159.33

ADJUSTMENT

INTEREST
CASH
CHECK
TOTAL

FRANKLIN TOWNSHIP   SOMERSET COUNTY

BLOCK NUMBER	LOT NUMBER	QUALIFICATION
552	1.01	

LOC:
DESC:
LOTS
DIMENSIONS: 350X110 .884 AC
BANK   Mortgage #   Tax Acct. #

EXPLANATION OF TAXES

DESCRIPTION	RATE PER $100	AMOUNT OF TAX
MUNICIPAL TAX:	.303	799.92
LESS PREV.TAX		3.18
BAL. OF MUNICIPAL TAX		796.74
NON-MUNICIPAL TAXES:		
COUNTY TAX	.265	699.60
COUNTY OPEN SPACE TAX	.030	79.20
DISTRICT SCHOOL TAX	1.158	3057.12
MUNICIPAL OPEN SPACE T	.050	132.00
FIRE DIST 1	.051	134.64

ASSESSED VALUATION INFORMATION

LAND	IMPROVEMENTS	TOTAL
264000	0	264000

EXEMPTIONS ►   NET TAXABLE VALUE ► 264000

NO. ...., NJ   07060

TOT NON-MUNICIPAL TAX	1.554	4102.56
LESS PREV.TAX		16.33
BAL NON-MUNICIPAL TAX		4086.23
TOTAL TAX		4882.97

2009 4TH QUARTER DUE NOV. 1, 2009	2441.48	2009 3RD QUARTER DUE AUG. 1, 2009	2441.49	BALANCE OF TAX DUE	4882.97

INFORMATION FOR TAXPAYERS

MAKE CHECK
PAYABLE TO: TOWNSHIP OF FRANKLIN

MAIL TO: COLLECTOR OF REVENUE
TOWNSHIP OF FRANKLIN
PO BOX 5059
SOMERSET, NJ 08875

SEE REVERSE SIDE FOR ADDITIONAL INFORMATION

REFER INQUIRIES REGARDING
COUNTY TAXES TO RICHARD WILLIAMS
(908) 231-7000.
SCHOOL TAXES TO EDWARD SETO
(732) 873-2400.
MUNICIPAL TAXES TO KENNETH DALY
(732) 873-2500.
3RD QUARTER TAX DUE BY 8/28/09 WITH
NO ADDITIONAL GRACE PERIOD.

2009 3rd & 4th Quarter Tax Bill

DISTRIBUTION OF TAXES

County Taxes	18.54%	$778.80
School Taxes	62.37%	$3057.12
Municipal Taxes	16.33%	$799.92
Other Taxes	2.76%	$134.64

STATE AID USED TO OFFSET LOCAL PROPERTY TAXES: The budgets of the government agencies
funded by this tax bill include State aid used to reduce property taxes. Based on the assessed
value, the amount of this State aid used to offset property taxes on this parcel equals: $921.36

FRANKLIN TOWNSHIP   **2009-4**
SOMERSET COUNTY
TAX COLLECTOR'S STUB - DETACH AND RETURN WITH YOUR CHECK
2009 4TH QUARTER TAX DUE NOVEMBER 1, 2009

BLOCK NUMBER	LOT NUMBER	QUALIFICATION	BANK CODE
552	1.01		

TAX ACCOUNT NUMBER	TAX BILL NUMBER	TAX AMOUNT BILLED	DUE NOVEMBER 1, 2009
	000288	►	2441.48

ADJUSTMENT
INTEREST
CASH
CHECK
TOTAL

FRANKLIN TOWNSHIP   **2009-3**
SOMERSET COUNTY
TAX COLLECTOR'S STUB - DETACH AND RETURN WITH YOUR CHECK
2009 3RD QUARTER TAX DUE AUGUST 1, 2009

BLOCK NUMBER	LOT NUMBER	QUALIFICATION	BANK CODE
552	1.01		

TAX ACCOUNT NUMBER	TAX BILL NUMBER	TAX AMOUNT BILLED	DUE AUGUST 1, 2009
	000288	►	2441.49

ADJUSTMENT
INTEREST
CASH
CHECK
TOTAL

www.ingramcontent.com/pod-product-compliance
Lightning Source LLC
Chambersburg PA
CBHW032016170526
45157CB00002B/721